Biology Workbook

NCEA Level 2

THIRD EDITION

David Blaker | Penny Daddy | Martin Hanson

Australia • Brazil • Mexico • Singapore • United Kingdom • United States

Biology Workbook NCEA Level 2
3rd Edition
David Blaker
Penny Daddy
Martin Hanson

Cover designer: Cheryl Smith, Macarn Design
Text designer: Cheryl Smith, Macarn Design
Production controller: Siew Han Ong

Any URLs contained in this publication were checked for currency during the production process. Note, however, that the publisher cannot vouch for the ongoing currency of URLs.

First published as Level 2 Biology Workbook in 2012

For product information and technology assistance,
in Australia call **1300 790 853**;
in New Zealand call **0800 449 725**

For permission to use material from this text or product, please email **aust.permissions@cengage.com**

National Library of New Zealand Cataloguing-in-Publication Data

978-0-17-037285-5

A catalogue record for this book is available from the National Library of New Zealand.

Cengage Learning Australia
Level 7, 80 Dorcas Street
South Melbourne, Victoria Australia 3205

Cengage Learning New Zealand
Unit 4B Rosedale Office Park
331 Rosedale Road, Albany, North Shore 0632, NZ

For learning solutions, visit **cengage.co.nz**

Printed in Australia by Ligare Pty Limited.
3 4 5 6 7 8 9 21 20 19 18 17

Contents

2.5 Genetic variation and change

2.6 Ecology

2.7 Gene expression

2.8 Microscope skills

Introduction

Using this resource

This workbook meets the needs of students doing NCEA Biology at Level 2, with all eight standards covered in detail. Each standard is subdivided into units, and each unit has two or three groups of self-check activities. Answers are provided at the back of the book. Information on NCEA requirements is placed at the end of each standard section. Hands-on practical investigations feature in five of the standards.

Internal

Each school has its own ways of dealing with internally assessed standards 2.1, 2.2, 2.3, 2.6 and 2.8. Some schools use individual student investigations and assignments for 2.3 and 2.6; others use internal tests. This book provides information to help students with a range of assessment types.

For each of the internal standards, dozens of different investigation topics are possible. It's not realistic to provide information to cover every possibility, but some topics are suggested together with outline information.

External

Achievement standards 2.4, 2.5 and 2.7 are assessed through external exams, so information here is provided in much greater detail. In addition, each of these sections has 10 exam-type questions that require longer answers. Half of these questions are accompanied by 'scaffold' information to help with the answering process. 'Scaffolding' is a very useful way of organising longer answers. The teacher CD contains answers to all 30 exam-type questions, together with model answers (exemplars) for some others.

A, M, E

This resource offers a depth of material for those aiming at any of the three grades: Achievement, Merit, Excellence. In some places, information more appropriate to Excellence grade is enclosed in framed boxes marked by E. This differentiation means that students aiming for Achievement grade could save time by skipping E-box material for that particular standard. Students aiming for Merit grade could consider using these E boxes.

In this resource and also in examinations, questions are not labelled A, M or E, because grades depend on answers, not on questions. However, NCEA questions provide clear clues as to what grade could potentially be earned. Most start with a verb, commonly *describe, explain, discuss*. Verbs like these give a strong indication of what a student needs to do to earn a particular grade.

Achievement: *describe, define, identify, list, match, name.*
Achievement with Merit: *explain, compare, classify, analyse, reason, summarise, sequence, apply.*
Achievement with Excellence: *discuss, evaluate, predict, design, plan, suggest, link, hypothesise.*

Practical investigation

NCEA Achievement Standard 91153

Carry out a practical investigation in a biology context, with supervision

2.1

Internally assessed, 4 credits.

Unit 1 | Guidelines

Any investigation for Biology 2.1 may be one of three types:

- **Fair-test experiments**. Examples: investigations involving osmosis, enzyme action, photosynthesis, bacterial growth, etc.
- **Pattern-seeking investigations**. Example: looking for patterns in the distribution of plants and/or invertebrate animals in a natural environment.
- **Modelling activities**. Example: using agar cubes to represent cells when investigating rates of diffusion.

Most of this unit relates to fair-test experiments. All investigations need to be supervised by teachers, who will also supply practical advice and background knowledge. Some investigations may involve group work, but planning stages and final reports must be done individually.

Question and hypothesis

Investigations usually begin with questions such as:

- What is the optimal fertiliser concentration for these plants?
- At what temperatures does yeast work fastest?

At optimal temperatures, yeast activity causes bread dough to rise.

Ideally, your question should link to something that already interests you. The next step is to use your question and your own knowledge to come up with a hypothesis that you could test. **Hypothesis** literally means 'idea under suggestion'. Hypotheses are best written as 'If ..., then ...' predictions that could be tested. Example: 'If we give these plants double the light intensity, then perhaps they will grow twice as fast.'

Fertiliser has little effect at early growth stages.

Using variables

Any living system is affected by a great many variables — which is a way of saying that life is endlessly complicated. One of biology's aims is to look for patterns, causes and links between

ISBN: 9780170372855

variables. In many cases, these links can be discovered by using the fair-test principle, which entails comparing two or more situations that are the same in almost every way.

An **independent variable (IV)** is the factor that you choose to alter. We say 'independent' because the choice is yours.

A **dependent variable (DV)** is the factor likely to be influenced by the IV. We say 'dependent' because it depends on the IV. The dependent variable gives your **results** (**data**). It is not correct to say that a DV is 'what is being measured'. In a good investigation, all variables should be measured.

Controlled variables (CVs) are all the other variables that could possibly affect your results. A CVs list depends on what is being investigated, and can include variables such as temperature, light intensity, chemical concentration, etc. In order to make comparisons as fair as possible, CVs need to be kept as constant (controlled) as possible in your trials. If these CVs are not the same for each trial, then the experiment is not valid; not properly set up.

Example:

Suppose you want to investigate how much heart rate changes when exercise time is increased.
The **independent variable** is the duration of exercise: 1, 2, 3, 5, 10 minutes, etc.
The **dependent variable** is heart rate, measured before and after exercise.
The **controlled variables**: same person(s), same type of exercise, same conditions.

Getting started

Start with a written plan. Even if you change your mind on a number of points, your plan provides guidance along the way. It could be as short as one page.

- Begin with a **question** you hope to answer. Write a **hypothesis** related to this.
- Write a **planned method** — what you intend to do. Identify IV and DV. Decide on units of measurement. The standard requires a range of four or more for the IV, such as four different temperatures. Make a list of CVs and suggest how you will keep them constant. Suggest how data will be collected for DV results. Suggest how many repeats you intend to do.
- Do **trial runs** to check for problems. Adjust your plan if necessary. If it is a group experiment, decide beforehand who does what.
- Experiment methods need to be **valid**, clearly based on fair-test principles.
- **Avoid bias**. Bias can arise from: (a) Different people making measurements (such as timing) in different ways; (b) Selecting results that you think will show what you were hoping to find.
- Use **correct units** for all variables. Example: temperature units are °C, not C.
- Single measurements can be misleading; a sample of one is never enough. **Multiple measurements** are more reliable than small samples. Outliers (extreme results) should not be averaged.
- Write **results straight away**, not on loose scraps of paper. Use photos and digital recordings where suitable.

ISBN: 9780170372855

Final report

After your investigation is complete, write it up as a proper scientific report. Recommended sub-headings:

1. **Question** or **Observation** that got your investigation started.
2. **Hypothesis**. Your aim is to test a hypothesis related to your initial question.
3. **Start plan**. Your notes should list all that was written down at the planning stage (above).
4. **Method**. What you actually did. Ideally written as a step-by-step summary. State if you adjusted your starting plan on any point, and why. Needs enough detail to enable anyone else to do the experiment.
5. **Results**. What you saw. Can include photos, observations, measurements, tables, graphs, calculations.
6. **Conclusion**. What the results tell you. A valid conclusion should link back to your original hypothesis, but in many situations the results may lead to no clear conclusion. Be honest about uncertainties. A conclusion is not a summary of results.
7. **Discussion**. This is your evaluation. Comment on possible sampling or measurement bias in your method. State what you did to make your method valid and your data reliable. State how well (or not) your results link in with other student results and also with theory.

Fair-test examples

There are hundreds of possible fair-test experiments — your teacher will provide guidance and direction. Also see the summary of NCEA guidelines at the end of Unit 2. Three are introduced here as starter questions.

- What range of soil pH will tomato seedlings tolerate?
- What foods do slaters prefer?
- What is the lowest concentration at which disinfectant X can sterilise Petri dishes?

Slaters (aka woodlice or isopods) are commonest in stacks of damp, partly decayed wood.

Activity A

1 Write a suitable hypothesis based on each of the 'fair-test' questions listed above.

a Tomatoes and pH

 ISBN: 9780170372855

b Slater food preferences

c Disinfectant X

2 Write the word(s) that match each definition or description in the table below. Select from this list: *hypothesis, conclusion, prediction, dependent variable (DV), questions, table, data, controlled variables (CVs), valid, independent variable (IV).*

a	The variable you choose to alter in an experiment	
b	The variable that provides results	
c	Need to be kept the same, to make comparisons fair	
d	Good method and measurements made without bias	
e	Information that provides results; includes measurements	
f	Means an idea that is 'under suggestion'	
g	A good starting point for any investigation	
h	What the results mean	
i	An organised way to record raw data	
j	What you think is likely to happen	

Pattern-seeking investigations

If you choose a pattern-seeking investigation, data collected for ecological topics could also be used for Biology 2.6. Your teacher will provide guidance and direction. There is a wide range of possible investigations. Three are introduced here as starter questions.

- Which animals predominate in each zone on that rocky shore?
- How do plants and insects differ in this undisturbed natural habitat, compared with the human-influenced habitat nearby?
- How do leaf sizes and shapes compare on the shaded and sunny sides of karaka trees?

Karaka. In some cases, leaf size is influenced by the intensity of light.

ISBN: 9780170372855

E

The nature of science

Science is a way of inquiry and investigation into the natural world. The main motivations: curiosity, plus putting knowledge to practical uses. After centuries of progress there are now many things we know for sure — but also great numbers of unanswered questions. Many facts are solidly proven but some areas of knowledge are much less certain.

Science investigates the unknown. Systematic processes include making observations and coming up with questions, followed by hypotheses that are then tested. Hypotheses may either be confirmed or else modified when new evidence is collected. However these logical steps are not the only ways in which knowledge advances: Einstein valued imagination and intuition, and asked highly original questions that at the time seemed un-answerable.

Most scientific advances are built on the combined efforts of teams of researchers doing similar work, often in different parts of the world. Results accumulate, followed by discussion on how best to interpret new findings.

Activity B

The following questions relate to an investigation on karaka leaf sizes (see page 9). Hypothesis: The leaves on a tree's shaded side grow larger in order to compensate for lower light levels.

During this investigation, many leaves need to be picked and measured. You can't pick every leaf on a tree, so take a sample. This must be done in a way that makes for a fair comparison between leaves on the sunny side of the tree (north facing) and the shaded side (south facing). Both samples of leaves must be picked at the same height, say 2 m above ground.

1 Apart from height, suggest two other ways in which both sampling methods need to be similar.

____

____

____

2 Suggest how you could avoid unintentionally picking big or small leaves, as this would cause bias in the results.

____

____

____

____

3 Suggest a suitable sample size (leaf numbers). ______________________________

 ISBN: 9780170372855

4 Make a sketch to show exactly what would be measured for each leaf.

5 Apart from leaf sizes and numbers, suggest two other variables that could usefully be measured, and how.

6 If you were to pick leaves from one side of a tree, and your work partner picks leaves from the other side, there would be unintentional bias. Explain what kind of bias could be caused by using two observers in this way.

ISBN: 9780170372855

Unit 2 | Case studies

The four experiments outlined here as case studies provide practice in planning, presenting data, and interpreting results. Activities B, C and D can also be done as part of the cell biology standard, 2.4.

Activity A | Nut energy

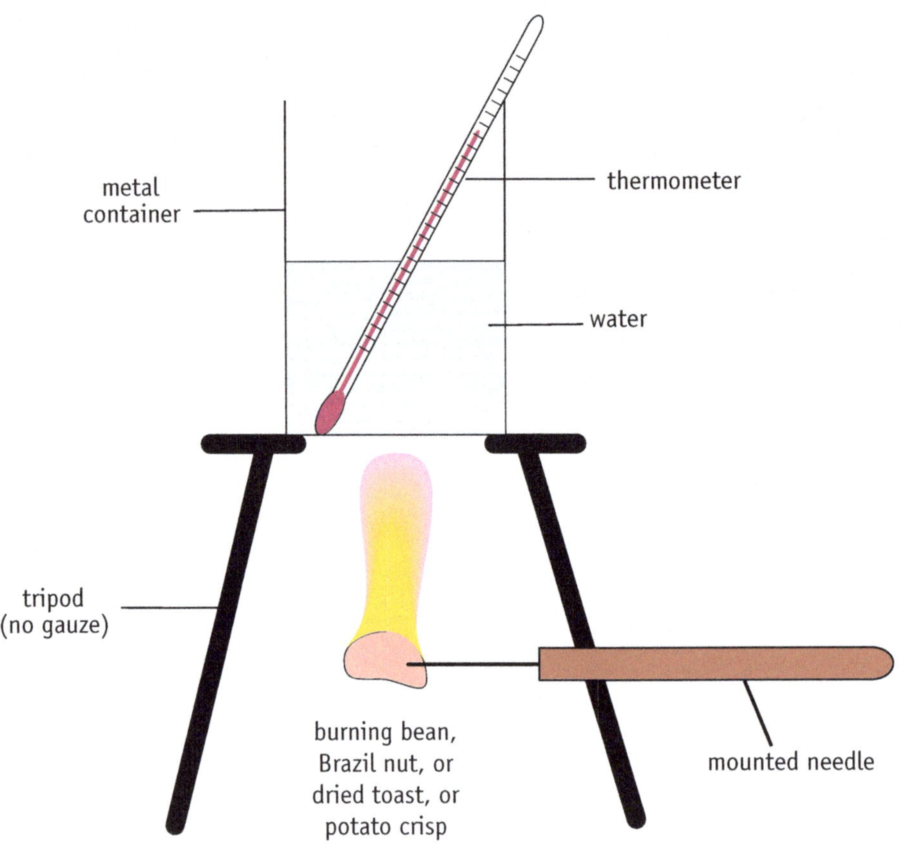

This simple experiment is used to compare the energy content of different foods. A piece of dried food is set alight in a Bunsen flame, and once the food has begun burning it is transferred beneath a container of cold water until the flame dies out. The temperature rise of the water gives a rough indication of how much chemical energy was in the food.

1 Suggest a hypothesis that could be tested using this technique. ______________________

____

____

2 Identify the independent variable (IV) in this experiment. ______________________

3 Suggest a suitable range for this IV. ______________________

____

 ISBN: 9780170372855

4 Identify the dependent variable (DV) in this experiment. ____________________

5 Identify a suitable unit of measurement for this DV. ____________________

6 If the aim is to compare three different foods, then all three tests need to have conditions (CVs) kept as similar as possible. List at least five variables that need to be controlled.

7 Evaluate the method by identifying at least one major defect in the experiment design as shown in the drawing. Suggest changes to the design that could make the results more valid.

Tables and graphs

Before you start any experiment, it's best to have a results table or spreadsheet ready and headed up. This avoids problems caused by having information on different bits of paper. Put your raw data directly into this table. Even if the results table is excellent, putting the figures into graph form will make patterns visually clearer. The following guidelines make a graph easier to read.

- **T.** Give it a descriptive **title.** 'Photosynthesis graph' is too vague.
- **A.** Label each **axis**. IV goes on the horizontal axis, DV (results) on the vertical axis. Put in the units, e.g. 'Time (minutes)', not just 'Time'.
- **D. Divide** each axis so it shows the full range of values. Write in value numbers spaced in even steps, e.g. 0, 10, 20, 30, 40.
- **P. Plot** the data, using **x** to mark each pair of numbers.
- **L.** Draw a **line** of best fit to show the trend. This could be a straight line or a curve.

The above guidelines apply mainly to line graphs. Spreadsheets will do some of this for you automatically, but you still need to make decisions on title, axis, line — and type of graph. Different guidelines apply to bar graphs, pie charts, etc; detailed suggestions on these are given in *Biology Workbook NCEA Level 3*.

ISBN: 9780170372855

1

Activity B | Photosynthesis

The table below gives the results of a class experiment aimed at measuring the rate of photosynthesis in a water plant. The rate at which oxygen bubbles were given off can be an indication of how fast photosynthesis is 'working'. The experiment was repeated five times at six different light intensities. (Standard classroom lighting is about 400 lux; full sun can be over 40,000 lux. The lux, symbol 'lx', is a unit of brightness.)

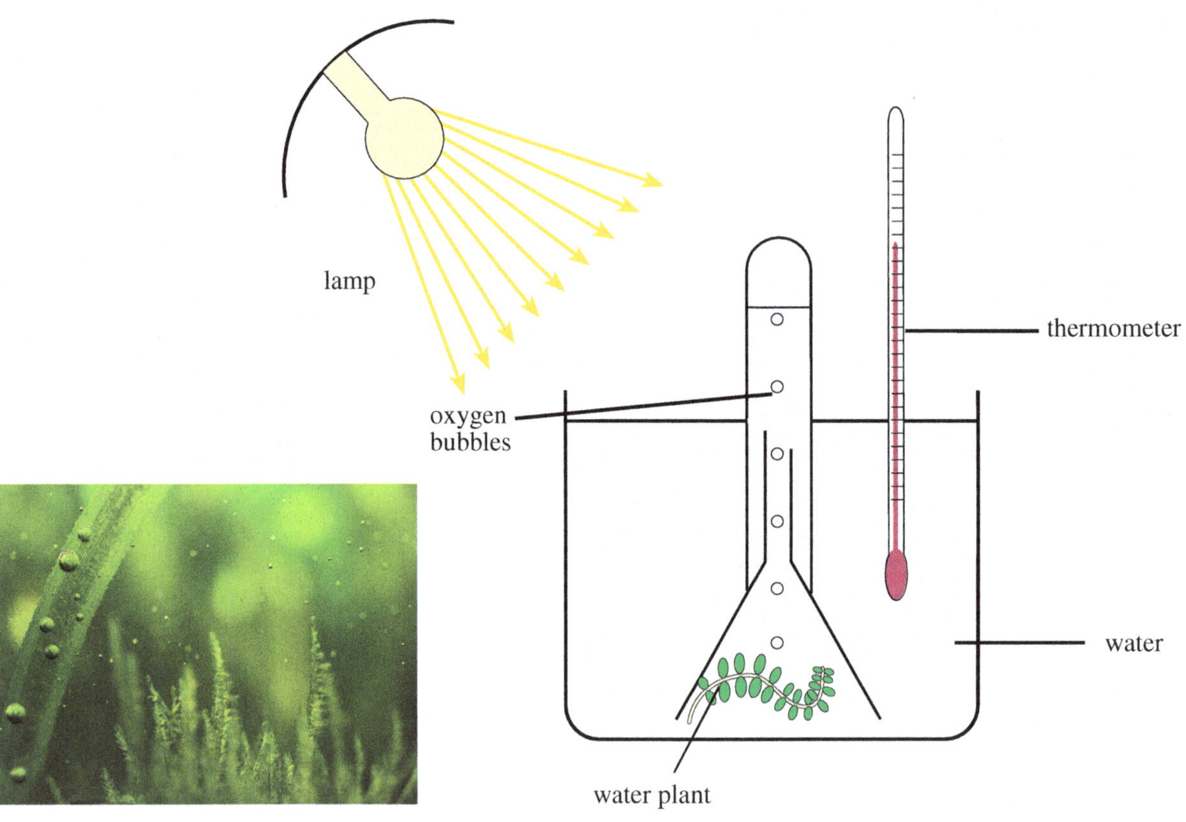

Light brightness (lx units)	Raw results: bubbles per minute	Average number of bubbles per minute
250	5, 4, 5, 3, 6	
500	21, 15, 17, 23, 18	
1200	43, 49, 59, 450, 49	
2000	62, 47, 57, 59, 61	
3000	59, 70, 64, 62, 62	
3500	69, 66, 60, 59, 62	

1 Calculate averages and put them in the column on the right.

 ISBN: 9780170372855

2 Which result would you discard? Justify why. Suggest how this error arose.

3 Use the blank graph below and the guidelines above to present these results in graph form.

4 Extrapolate the graph to predict the likely bubble rate at 5000 lux.

5 Extrapolate the graph to predict at what brightness the bubble rate is likely to become zero.

6 Identify the independent variable in this experiment, and state what range was used.

7 Suggest three variables that would need to be controlled (kept constant) throughout this experiment.

8 Link your results graph to information on photosynthesis and limiting factors. (See pages 52 and 81.)

ISBN: 9780170372855

1

Activity C | Osmosis

Background

When two different solutions are separated by a semi-permeable membrane (spm), water tends to diffuse from the more dilute solution to the more concentrated solution (D→C). This movement is known as **osmosis**. 'Concentrated' and 'dilute' refers to the amount of dissolved chemicals such as salts and sugars. A solution that has 20% sugar is 80% water; one with 5% sugar has 95% water. This means that osmosis moves water from where there is more water to where there is less water. Osmosis affects all plant and animal cells that are in direct contact with water. Seawater is about 3.5% salt; most cells have much lower concentration. Any cell membrane acts as an spm.

It is very difficult to work with single cells, so in this experiment whole pieces of kumara are used as a cell model, each piece consisting of thousands of similar cells. The results could tell us about the solutions inside kumara cells. (Instead of kumara, potato or carrot could be used.)

Question and Aim

The aim is to try to answer the question: At what point does osmotic inflow balance outflow?

Hypothesis

Write a hypothesis that you could test in this experiment.

____

____

Method

- Prepare NaCl solutions with a range of different concentrations: 0%, 0.5%, 1%, 2%, 3%, 4%. If your chemical knowledge is sufficient, you could instead make solutions ranging from 0 M to 0.7 M.
- Place each solution in a separate large test tube or small beaker. Next, use a cork borer to cut kumara 'flesh' into neat same-size cylinders. Alternatively, cut it into cubes with sides of 10 mm.
- Weigh each piece to 0.1 g; record its weight. Don't get them mixed up. Put three kumara pieces in each solution. Leave for 24 hours.
- Remove pieces, lightly blot each one dry; immediately measure the 'after' weight of each. Enter these weights in the results table below.

1 Identify the independent variable. _______________

2 Suggest what units will be used to measure the IV. _______________

3 Identify the dependent variable. _______________

4 Suggest what units could be used to measure the DV. _______________

5 List at least five controlled variables (CVs) that will need to be kept as constant as possible for all the kumara pieces.

____

____

____

ISBN: 9780170372855

6 Suggest how many pieces you will need for each solution, and how many in total.

7 Suggest how to prevent pieces and their weights becoming mixed up.

8 Suggest how to prevent the cut pieces becoming dry in air and losing weight.

9 Using the planning information on the previous page, plus your own decisions on the above questions, draw a flow chart of your whole method. Number and briefly describe each step in the overall process.

Results

Write results in immediately after weighings are done. Depending on how many pieces are used, you may need to add rows.

Piece number	Solution conc.	Before (g)	After (g)	Gain or Loss (g)	% G or L

ISBN: 9780170372855

1

Add information to the blank graph (title, axes labelled, units and divisions correctly shown). Place '% Gain/Loss' on the vertical axis, with some way of showing weight gains up to 20% and weight losses up to 20%. Do this in pencil so that it can be changed to suit your actual results when they arrive.

Conclusion (refer to Unit 1)

Discussion (refer to Unit 1)

ISBN: 9780170372855

Activity D | Enzyme activity at different temperatures

Background

When temperature increases, enzymes work faster. However, above a certain temperature enzyme molecules start to become denatured (damaged by heat), and when the temperature gets too high they are irreversibly destroyed. Each kind of enzyme has an optimal (ideal) temperature. In the human body, many enzymes function fastest at 37 °C; elsewhere, some enzymes function best at temperatures as high as 60 °C. One enzyme occurring in most living cells is **catalase**, which speeds up the breakdown of poisonous hydrogen peroxide to water and oxygen.

$$2H_2O_2 \rightarrow 2H_2O + O_2$$

One way of measuring catalase activity is to use a suspension of live yeast cells, then measure the rate at which oxygen bubbles are given off. Once the setup is working well, trials can easily be done at different temperatures. Use a fresh lot of hydrogen peroxide and yeast for each trial.

1 Define 'enzyme'.

2 State which exact temperatures you would choose for the above experiment.

3 Suggest how you would change the enzyme's temperature.

4 Write your own plan for the above experiment. Use these four headings as a guide: Question; Hypothesis; Full method plan; Results table. First read the A, M, E criteria on page 20, then aim for Excellence in your method plan. Continue on a separate page.

ISBN: 9780170372855

Biology 2.1 Practical investigation

AS 91153 Carry out a practical investigation in a biology context, with supervision
Internally assessed, 4 credits

Achievement	Achievement with Merit	Achievement with Excellence
Carry out a practical investigation in a biology context, with supervision.	Carry out an in-depth practical investigation in a biology context, with supervision.	Carry out a comprehensive practical investigation in a biology context, with supervision.

Achievement

'Carry out a practical investigation …' involves:

- developing a statement of the purpose written as a hypothesis linked to a scientific concept or idea
- using a method that describes:
  - for a fair test: a range for the independent variable, the measurement of the dependent variable and the control of some other key variables
  - for a pattern-seeking or modelling activity: the data that will be collected, range of data/samples, and consideration of some other key factors
- collecting, recording, and processing data relevant to the purpose of the investigation
- interpreting and reporting on the findings
- reaching a conclusion based on the processed data which is relevant to the purpose of the investigation
- identifying and including relevant findings from another source.

Achievement with Merit

'Carry out an in-depth practical investigation …' involves:

- using a method that describes:
  - for a fair test: a valid range for the independent variable, the valid measurement of the dependent variable and the control of other key variables with consideration of factors such as sampling bias and sources of errors
  - for a pattern-seeking or modelling activity: a valid collection of data with consideration of factors such as sampling bias and sources of errors
- collecting, recording, and processing data which enables a trend or pattern (or the absence of a trend or pattern) to be determined
- reaching a conclusion based on the processed data which is relevant to the purpose of the investigation
- a discussion of the biological ideas relating to the investigation that is based on the student's findings and those from other source(s).

Achievement with Excellence

'Carry out a comprehensive practical investigation …' involves justification of the choices made during the sound investigation, i.e. evaluating the validity of the method or reliability of the data and explaining the conclusion in terms of the biology ideas relevant to the investigation.

Notes (Summary only. For further details, visit the NCEA website.)

1 A 'practical investigation' is an activity covering the complete investigation process: planning and carrying out the investigation, collecting primary data, processing and interpreting data, and reporting on the investigation. Students may make changes to their initial method as they work through the investigation.
2 Assessment against this standard may be based on a stand-alone or an individual investigation that can contribute findings to a larger group or class investigation.
3 The nature of the investigation could be the manipulation of variables (fair test), the investigation of a pattern or relationship or the use of models.
4 It is intended that this investigation be carried out with supervision.

ISBN: 9780170372855

Analysing information

NCEA Achievement Standard 91154

Analyse the biological validity of information presented to the public

2.2

Internally assessed, 3 credits.

Unit 1 | Guidelines

Many socio-scientific issues affect daily life and in many cases we need to make our own decisions on these. Below are three examples of science-based issues that society faces. We could leave decisions on all such matters to technical experts and politicians, but outcomes will be better if everyone is well informed and able to evaluate evidence.

Examples:

- **Antibiotics**. Why are they are becoming less effective? Should we cut back on their use?
- **Immunisation**. How much do the advantages outweigh the risks? How good is the evidence on each side of the debate?

More contexts and topics are listed on the next page. Whichever gets chosen, it needs to be one that directly affects New Zealand or the nearby Pacific region.

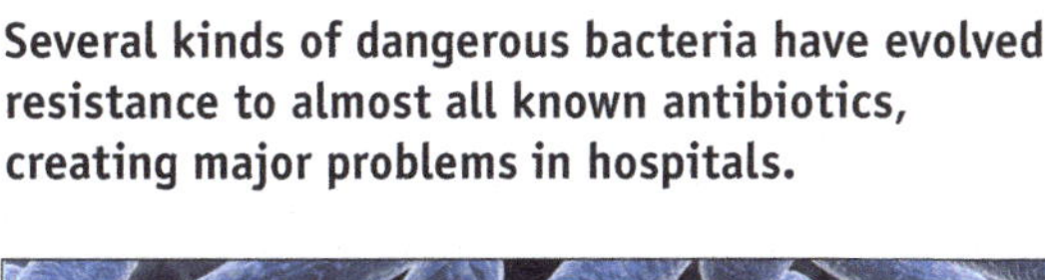

Several kinds of dangerous bacteria have evolved resistance to almost all known antibiotics, creating major problems in hospitals.

If people 150 years ago had thought more carefully about the outcomes, possums would never have been introduced to New Zealand.

ISBN: 9780170372855

2

Selecting a topic

This chapter is designed to help you with internally assessed Biology Achievement Standard 2.2, requirements for which are given at the end of the next unit.

Your teacher may select a topic for you and provide a list of websites and articles for you to analyse. Alternatively, you may be allowed to select a topic — in which case the teacher needs to approve your selection. The following criteria apply.

- Information should be from a range of different sources; at least three.
- Resources — whether print or internet — should be three different genres (types).
- The topic must be mainly biological. Avoid topics like abortion and euthanasia, which are moral issues, not scientific ones.
- The topic must be one that directly affects New Zealand or the Pacific region.
- Students need a good knowledge base. Example: to deal with a topic on toxins, you will need a reasonable knowledge of chemistry.
- A narrower topic is often easier to handle than a wide one. Example: instead of 'The possum problem', consider narrowing the topic to something like 'What are the advantages and drawbacks of using 1080 poison?'
- To help provide focus, it may help to start with questions like the one above.

Some biological topics and contexts

Coral reef bleaching and decline. Conservation of the New Zealand falcon. Biofuels vs diesel. Maui's dolphin and bans on set-net fishing. Stoats: damage and control. Acne-treatment creams: how effective and safe are they? To what extent is farm run-off affecting the water quality of lakes and rivers? Cage farming vs free-range farming for pigs. Why are antibiotics becoming less effective, and how bad is the situation? Organic farming methods: do they improve soil quality? Mangrove spread, its causes and consequences. How much tanning is safe? Can GM eliminate human diseases, and if so, which ones? Kiwi decline and conservation. The pros and cons of allowing mining in national parks. Music and hearing loss: what is the evidence? Overfishing. Marine reserves. GM plants and the spread of 'super weeds'. Herbicide uses and risks. Tiger conservation: the role of zoos. Vanishing bees. Using 1080 poison: risks and advantages. Fluoride, water safety, and tooth decay. The effect of nature tourism on endangered species. The role of farming in conservation. Mobile phones: the evidence for safety and long-term harm. The effect of cannabis on brain development. The role of food in diabetes. Sustainable forestry. Fishing and the spread of problem water plants. Immunisation: the evidence. The advantages and disadvantages of open-access conservation islands compared with closed-access islands. Different kinds of wild pig control in New Zealand wilderness areas. Biological farming vs high-input farming. Climate change and its effects on farming practices in New Zealand. Tanalised-treated timbers: advantages and problems.

 ISBN: 9780170372855

Method in science

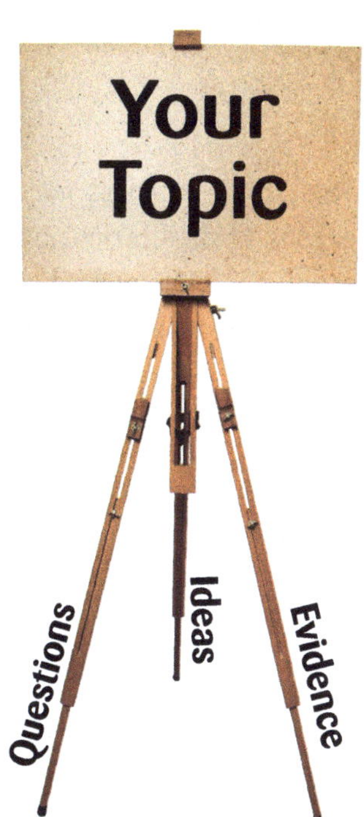

Each socio-scientific issue is different, but it helps to have a big picture of how science works. Science, including biology, stands on three legs. These 'legs' are:

- questions (based on curiosity and a desire to find out)
- ideas (including hypothesis and theory)
- evidence (information, including reliable facts).

How do we know what's true and what's not? Is everything just a matter of opinion? Fortunately, the scientific method helps you sort out good ideas from bad ones. If plenty of solid evidence exists, an idea will survive. If evidence is weak, the idea may be thrown out.

When analysing information on a topic, we need to judge the validity of what is being presented as fact. We can reduce this to simple questions:

- Is the information scientifically **accurate or inaccurate**?
- Is the information **biased or not**?

Three stages

When you have a suitable topic your remaining tasks are:

- collecting information
- processing this information
- final presentation.

Except for teacher guidance, all three stages must be completed individually.

Collecting

Your teacher may provide a range of suitable print and internet articles for you. Any that you locate for yourself will need to be chosen carefully because they will become the basis for the processing stage. These articles will become your 'research material' and you need to keep them so your teacher can check the information and also help you clarify your ideas as you go along. You may have to look at dozens of articles before selecting a core group that will be your research material. Whether from print media or from websites, the research material needs to have these features:

- Three different genres (types) aimed at different audiences. Some examples: health education, advertising, general magazines, newspaper articles, science magazines, government statistics, original 'source material', research reports, TV documentaries.
- Represent a range of opinions with opposing viewpoints. This will enable you to consider two or more sides to a topic.

2

Processing

This stage consists mainly of analysing your selected research material articles, and judging to what extent their information is accurate and without bias. For more guidance, see the section titled 'Good science, bad science, bias' (page 25). For practice in processing and analysing, see the articles in Unit 2. As you process information, consider these questions:

- Are there two sides to the story? More than two sides?
- How can we be sure the information is correct? What can you rely on?
- Is the evidence weak or strong?
- What are the consequences of good science evidence?
- What are the consequences of inaccuracy and misinformation?
- Which items in this article are more important, and which less important?
- It is not enough to just make general statements like 'probably accurate', or 'looks suspicious'. You need to refer to specific parts of the article(s), and explain in what way they are accurate or inaccurate, biased or not.

Presentation

This can involve a written portfolio (collection) covering each resource individually, or else a single report on all resources collectively. If your teacher approves, other possible methods of presentation could be a webpage, a talk illustrated by a PowerPoint presentation, or a video. However, your report will be assessed on your processing and analysis of the information, not on its presentation.

Reference list
Whether aiming for A, M or E, you must include a reference list of all the resources you used, so that anyone reading your report can locate the sources again. Each reference is to include: author name, date of publication, title, publisher name, web address.

Make a note of these requirements at the start:

Number of class periods available	
Final hand-in date	
Teacher's suggestions on presentation type and length	

Activity A

In the table at the top of page 25, write the word(s) that match each of the definitions or descriptions below. Select from this list: *analyse, bias, evidence, hypothesis, objective, proof, theory, valid, vested interest*.

Use a dictionary if you are uncertain about some meanings, as it's important to use the above words correctly in your final report.

ISBN: 9780170372855

1	Facts or signs showing that something exists or is true	
2	Strong evidence that is totally convincing, or almost totally	
3	Any suggestion that can be a starting point for investigation	
4	Argument or evidence or experiment that is reasonable and well justified	
5	In science, maths and economics, this means a big idea that explains many facts	
6	To break something down into its constituent parts	
7	To influence attitudes, choices, decisions	
8	When used as an adjective, this means facts not influenced by personal opinions	
9	A person or organisation having — and often hiding — a strong reason or personal agenda for wanting something to happen because they will gain from it	

Good science, bad science, bias

How can we filter out poor-quality information? This is a particular problem of the internet, as any search turns up much material that is useless or misleading. The list below suggests 10 ways of detecting the difference between valid information on the one hand and inaccurate information and bias on the other.

Questions 1–6 relate mainly to **bias**.
Questions 7–9 relate to the **quality** of information; whether it is valid or not.
Question 10 relates to the **quantity** of recent scientific information.

1. Does it hold **hidden bias**? Open bias and advertising are easier to deal with.
2. Is the writer's **name and identity hidden**? Writers who state their name and organisation are more likely to be reliable. Most university research is objective.
3. Are there **vested interests**? Be cautious about information given out by those who stand to gain money or power.
4. Does it **ignore alternative points of view**? Bias and media spin are often revealed in what an article does not say.
5. Does it involve unreasonably **emotive language and personal attacks**? The 'comments' sections of some websites are particularly bad in this way.
6. Is it promoting a **conspiracy plot**? Most conspiracy ideas are nonsense. Example: 'They are secretly poisoning our water with chemicals.'
7. **Is it tricky**? Common tricks: **a** Quoting statements out of context. **b** Opinions mixed up with facts. In good reporting, facts and opinions are kept separate. **c** Creating doubt by claiming there is too much uncertainty to decide. **d** Using a few points of evidence to back up a pre-decided point of view, and ignoring the overall weight of evidence. **e** Writing for the general public then confusing readers with technical information.

8 Does it use **false experts**? The question becomes: expert on what? Example: regarding conservation, ecologists rank higher than nuclear scientists.

9 Is it based on **weak evidence**? Some research is weak, some is strong. The most useful reports are probably recent, and also peer reviewed. (**Peer reviewed** means that the report has been checked and approved by a panel of specialists in the same branch of science.)

10 Are **a few selected items of evidence presented** as though they are the whole story? On many subjects there is a wide range of evidence. Often a research report will come up with different evidence to previous reports on the same subject, but one report is seldom convincing. It takes a collection of strong evidence to build a convincing case.

2

E

Internet problems

The internet presents special problems to anyone searching for reliable information. Many websites are excellent, but many other sites that look like useful sources turn out to be false, unreliable, biased and misleading. Boundaries between fact and opinion may be blurred.

Your best protections against these pitfalls are: **a** an attitude of reasoned scepticism; **b** making sure that a website is linked to an organisation whose purposes and credentials you can check on.

However, even a scientific-sounding title can be misleading. Example: the New Zealand Climate Science Coalition website claims to 'represent accurately, and without prejudice, facts regarding climate change'. In reality, all of its links are to the small fraction of science reports that cast doubt on climate change.

Another common problem is that some information is reliable, but much too technical and written for experts only.

Activity B

Choose any one example of 'bad science' that you have come across in the media.

Summarise the situation in a few lines, then explain why you consider it 'bad', such as having biased and/or poor quality information.

ISBN: 9780170372855

'It ain't what you don't know that gets you into trouble. It's what you know for sure that just ain't so.'

— saying attributed to Mark Twain

'We don't have the money, so we've got to think.'

— Ernest Rutherford

2

Economy vs environment?

When it comes to environmental matters, does the economy need to be balanced against the environment? This view is summed up in diagram A, which shows economy and environment as different and opposed. However, we could also view the economy as a subset that depends entirely on a much greater natural environment (diagram B). If natural environments are neglected or degraded or polluted, then economic damage is likely to follow. This would be in addition to the intrinsic value of healthy diverse natural environments.

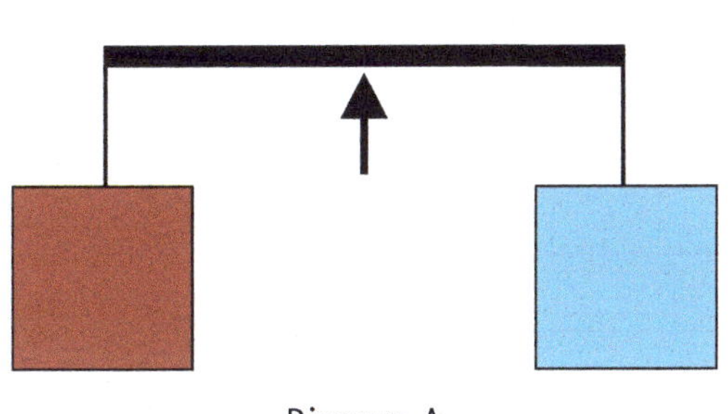

Diagram A

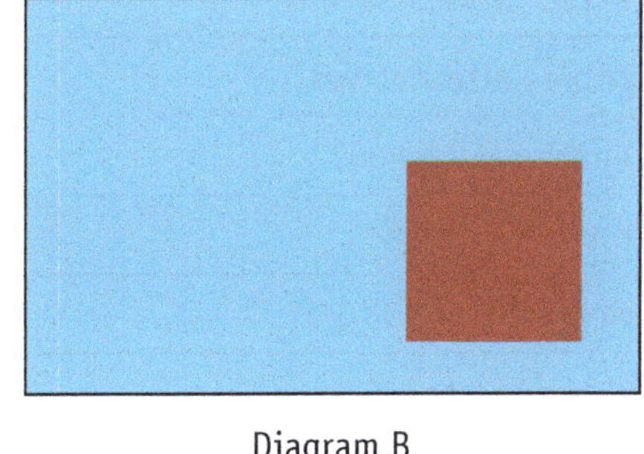

Diagram B

ISBN: 9780170372855

Unit 2 | Fluoridation

2

Of the many socio-scientific issues listed in the previous unit, one is dealt with in more detail here: fluoridation. The element fluorine, in the form of fluoride ions (F^-), occurs naturally in drinking water in many places in the world. It has long been known that people who drink this water seldom have tooth decay. For decades now it has been common practice to add small amounts of fluoride to water supplies, at concentrations of less than 1 ppm (part per million). At much higher concentrations fluoride is poisonous, which causes some people to object to fluoridation.

Below are extracts from three different resources on the same subject. Each is just part of a longer article, so full analysis is not possible here: it is a practice activity. Read all three carefully before attempting activities A and B.

Activity A

Give each resource a score out of 5 based on the 'good science, bad science, bias' list from the previous unit. On each line award 5 for maximum bad-science score, 0 for no bad-science features, n/a if it does not apply. Low scores are best.

	Bad science feature	Resource 1	Resource 2	Resource 3
1	Hidden bias?			
2	Name not revealed?			
3	Vested interests?			
4	Ignores alternatives?			
5	Emotive language?			
6	Conspiracy idea?			
7	Tricky?			
8	False experts?			
9	Outdated evidence?			
10	Selective evidence, or unreliable?			
	Total score			

Activity B

In the spaces provided below each resource:

- make notes on who the intended 'audience' is, and what genre resource it seems to be.
- make notes referring to particular features of bias, and your reasons for deciding that it is biased and/or from a vested interest.
- regarding the printed article itself, make notes indicating accurate and inaccurate items of information.
- correct any mistakes; note what appear to be dubious 'facts'.
- decide which article you rank highest for accuracy and validity and lack of bias, and which lowest. Give reasons for your rankings.

ISBN: 9780170372855

Resource 1

American Dental Association

www.ada.org/fluoride.aspx

Community water fluoridation is the single most effective public health measure to prevent tooth decay. The Center for Disease Control and Prevention has proclaimed community water fluoridation as one of 10 great public health achievements of the 20th century. Approximately 72.4% of the U.S. population served by public water systems receive the benefit of optimally fluoridated water.

Fluoridation of community water supplies is simply the adjustment of the existing, naturally occurring fluoride levels in drinking water to an optimal fluoride level recommended by the U.S. Public Health Service (0.7–1.2 parts per million) for the prevention of tooth decay. Water that has been fortified with fluoride is similar to fortifying milk with Vitamin D, table salt with iodine, and bread and cereals with folic acid.

Studies conducted throughout the past 65 years have consistently shown that fluoridation of community water supplies is safe and effective in preventing dental decay in both children and adults. Simply by drinking water, children and adults can benefit from fluoridation's cavity protection whether they are at home, work or school.

Today, studies prove water fluoridation continues to be effective in reducing tooth decay by 20–40%. Fluoridation is one public health program that actually saves money. An individual can have a lifetime of fluoridated water for less than the cost of one dental filling.

The American Dental Association continues to endorse fluoridation of community water supplies as safe and effective for preventing tooth decay. This support has been the Association's position since policy was first adopted in 1950. The ADA's policies regarding community water fluoridation are based on the overwhelming weight of peer-reviewed, credible scientific evidence. The ADA, along with state and local dental societies, continues to work with federal, state and local agencies to increase the number of communities benefiting from water fluoridation.

ISBN: 9780170372855

2

Resource 2

New Zealand Herald, 31 March 2013

Is adding fluoride to water a cheap way to help prevent tooth decay, or a breach of our personal rights?

By Chloe Johnson

Fluoride is a natural element in our water that helps strengthen teeth and prevent decay — but the levels at which it naturally occurs are not enough to fully protect people's teeth. In 1954, New Zealand became the second country in the world to introduce fluoride to some of its water supplies. Hastings was the first region to copy the Americans, who discovered a lack of fluoridation contributed to higher rates of tooth decay.

More than 84 municipal water supplies are now topped up with manufactured fluoride called hydrofluorosilicic acid, sodium fluoride or silicofluoride. About 52 per cent of New Zealanders currently have access to fluoridated water through council supplies and food. The New Zealand Drinking-Water Standards recommend optimum fluoride levels of 0.7 to one part per million in tap water.

Over the past 60 years, scientific studies argued about the health effects of fluoridated water. The debate has been a factor in convincing some councils, including Waipukurau and Northland, to stop adding fluoride. Others, like the towns of Waverley and Patea in South Taranaki, have been persuaded to introduce fluoride.

Advocates say adding fluoride to the water is a safe and cost-effective way to reduce tooth decay, particularly in low socio-economic groups.

Opponents claim it is a poison that triggers adverse health effects, not to mention a gross breach of individual rights. Those opponents have been derided as paranoid, tinfoil hat-wearers in the past — but now they're back with a glint in their eyes.

ISBN: 9780170372855

Resource 3

Fluoride safety

By Dr Wilson

www.drlwilson.com/articles/fluoridation.htm, August 2013

Fluorine is a highly toxic element. Proponents say it is a nutrient. Scientists are mixed on this point. If it is a necessary nutrient, only a trace amount is needed in the body. We get this from foods such as tea, and small amounts are found naturally in most drinking water.

Sodium fluoride, the chemical used in water fluoridation, is a cumulative toxin. It is sold as rat poison, used in pesticides, and is the active ingredient in sarin nerve gas. Fluoride tablets require a prescription, unlike any other nutrient mineral. All fluoride toothpaste comes with a warning label. The label states 'Keep out of the reach of children under 6. If you swallow more than used for brushing, seek professional assistance or contact a poison control center immediately.' This warning applies to anything greater than a pea-sized drop of toothpaste on your brush.

Fluoride is considered one of the worst, if not the worst airborne pollutant, responsible for decimating fish and wildlife populations. The United States is one of 22 nations that signed a treaty promising not to dump fluorides into the oceans, lakes or rivers.

In the doses that fluoridated water provides, fluoride is associated with higher rates of birth defects, cancer and immunosuppression, lower IQ of children, dental and skeletal fluorosis, and neurological problems. It has also been shown to cause increased bone fractures, cataracts and infant mortality, and some 20 other health effects. A study in Brain Research, vol. 784:1998 showed that fluoride in the water fed to rats increased the absorption of aluminum into the rats' brains, causing alterations in the brains similar to Alzheimer's Disease. Also, fluoride is highly corrosive. Several studies in Massachusetts and elsewhere found higher levels of lead in the drinking water in fluoridated areas. This is most likely due to corrosion of lead pipe joints as a result of the corrosive chemical.

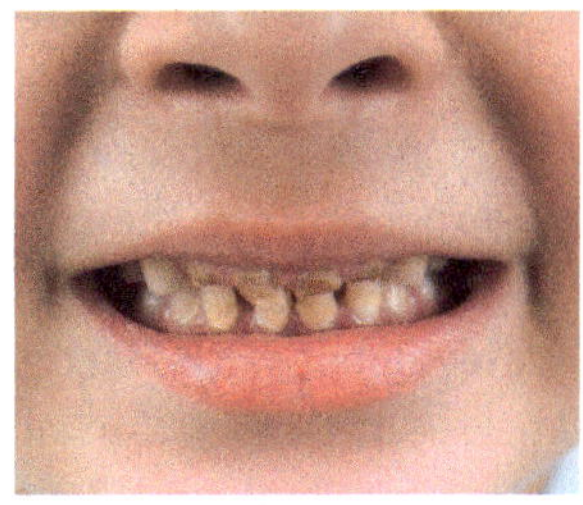

Thefluorideitselfisn'ttheonlyproblem.Thechemical used to fluoridate is not pure. Hydrofluorosilicic acid and sodium fluoride are industrial wastes, by-products of the phosphate fertilizer industry. This was challenged by the fluoride promoters in Ohio, but later they were forced to admit this is the truth. They contain traces of lead, arsenic, mercury, kerosene, naphtha, and other pollutants from the smokestack scrubbers of phosphate factories. They also contain radioactive elements.

ISBN: 9780170372855

Biology 2.2 Analysing information

AS 91154 Analyse the biological validity of information presented to the public
Internally assessed, 3 credits

Achievement	Achievement with Merit	Achievement with Excellence
Analyse the biological validity of information presented to the public.	Analyse in-depth the biological validity of information presented to the public.	Comprehensively analyse the biological validity of information presented to the public.

Achievement
'Analyse the biological validity ...' involves:
- recognising and describing biological features in the information and identifying them as accurate, inaccurate or biased using biological knowledge. Recognising inaccuracies may be demonstrated by making corrections to inaccurate biological features
- identifying the purpose of the information (e.g. who produced it and the intended audience).

Achievement with Merit
'Analyse in-depth ...' involves giving reasons why or how:
- each biological feature is accurate or inaccurate, or contains bias
- inaccuracies and/or bias may have consequences or impacts for the public
- vested interest is conveyed in the information.

Achievement with Excellence
'Comprehensively analyse ...' involves:
- prioritising, with reasons, aspects of the information in relation to their significance in the context
- evaluating the overall impact of the article on the public, based on bias and the balance of accurate and inaccurate features.

Notes
1 'Biological validity' refers to scientifically accurate information that is used in an unbiased way to convey a biological idea.
2 Students need to deal with at least three different kinds of resource, ideally ones having different purposes and intended for different audiences.
3 'Prioritising' refers to identifying why some items of information are more important than others.

 ISBN: 9780170372855

Plant and animal adaptations

NCEA Achievement Standard 91155

Demonstrate understanding of adaptation of plants or animals to their way of life

Internally assessed, 3 credits.

Unit 1 | Background

Every plant and animal is adapted to its own particular niche (its way of life) and habitat (where it lives). Whatever its niche and habitat, every living thing has to deal with the same four problems. All need to:

- get enough energy and food
- avoid being eaten
- deal with physical problems like dehydration, cold, heat, wind
- reproduce successfully.

Although the basic problems are the same for all plants and animals, we find an amazing variety of different features 'answering' the same problems. These features (or mechanisms) are known as **adaptations**. The function of a particular adaptation is known as its **adaptive advantage**.

In biology, adaptation does not mean 'change'. Adaptations are genetically inherited features that are encoded within DNA and passed from one generation to the next. (If you play a lot of sport, for example, your bones will become stronger, but this will not change your DNA, so we can't use the word 'adaptation'. The word for this kind of change is 'acclimation'.) Some adaptations involve structures, others mainly involve behaviour and/or biochemistry.

Karearea (New Zealand falcons) are specialised for catching small birds at high speed. Adaptations include having superb vision, fast reflexes, sharp claws for killing, and a hooked beak for tearing flesh. However, no adaptation is perfect: there are always limitations, and falcon prey often escape. If falcons were perfectly adapted, they would become abundant and eventually overeat their food supply.

Example: Weta

Habitat: forest trees with crevices to hide in.
Niche: eat plants; nocturnal.

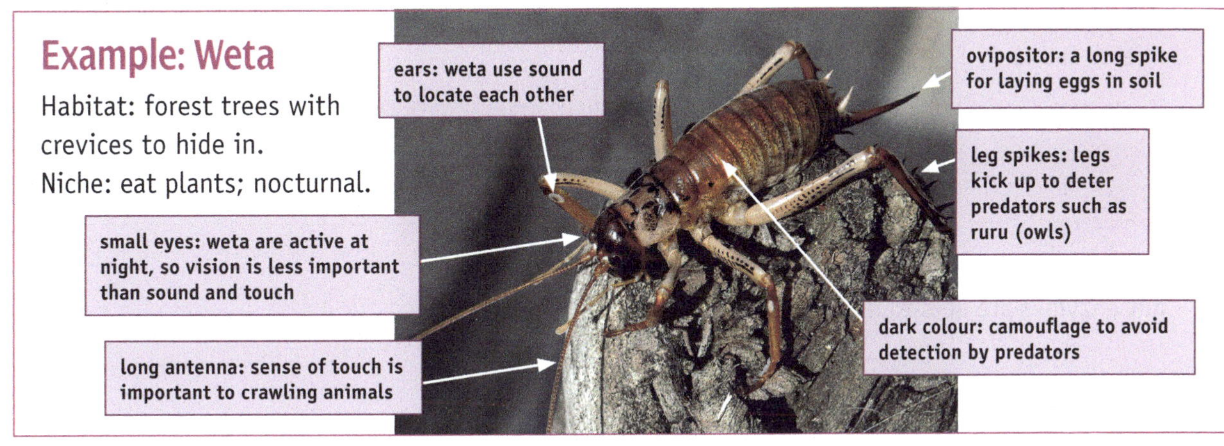

ISBN: 9780170372855

3

Activity A

Listed below are some adaptations of two very different organisms in different taxons (groups). Each is adapted to its particular niche and habitat. For each feature, briefly explain its **adaptive advantage** — why it helps the organism succeed in its niche.

Cactus

Habitat: hot dry deserts, only occasional rain. Niche: photosynthesis, uses insects for pollination.

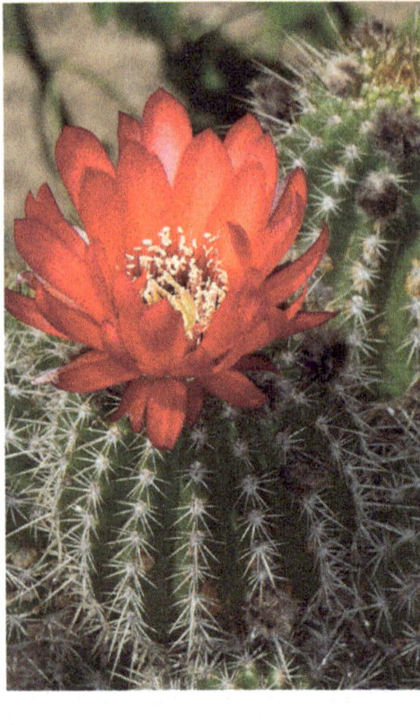

1 spiky thorns ______________________

2 stem for photosynthesis (has no green leaves)

____

3 flowers only after rain ______________________

4 waxy surface ______________________

5 the stem epidermis is almost transparent

____

6 stem swells after rain ______________________

Adult mosquito

Niche: bloodsucking parasite.

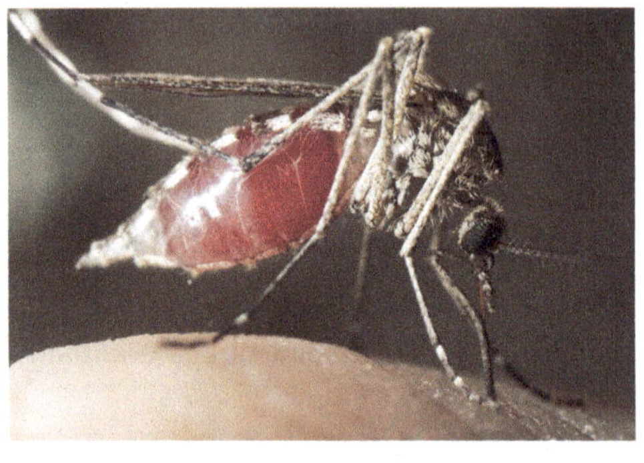

1 antenna sensitive to odour and temperature

____

2 big eyes, good eyesight

____

3 sharp proboscis ______________________

4 injects saliva ______________________

5 exoskeleton between body segments is flexible

____

6 gut enzymes are able to digest protein ______________________

ISBN: 9780170372855

How and why?

Describing an adaptation is a start, but it also needs to be explained. It helps to recognise that two different kinds of question apply in these situations: 'How?' and 'Why?'

Example: insect-eating bats use sonar to catch their prey

How? (Explain the mechanism.) Bats produce and aim high-frequency sounds, and react to echoes from moths, etc. These are **adaptations**.

Why? (Explain the function.) Bats hunt at night, when vision is of little use. Sonar gives them greater hunting success. These are **adaptive advantages** of sonar.

E

Thinking about questions

Science is full of questions: What? Why? How? When? etc. In the case of plant and animal adaptations, questions fall into three main categories. When thinking about the categories described below, it helps to have a particular adaptation in mind, whether bat sonar, a spider's web, or whale migration.

Mechanism. This generally involves **how** questions such as: How does it work? What is going on inside? What trigger sets it off? What chemical process is involved?

Function. This generally involves **why** and **what for** questions such as: Why does it help the animal survive? Why does it increase reproductive success? What is it for? Why is it different from species X? These questions relate to **adaptive advantages**.

Origin. How and when did it evolve? How did it get to be this way?

'Mechanism' questions exist in all branches of science; but function and origin questions are meaningful only in biology. In chemistry and physics, you can't really question what atoms are for, or how energy began.

ISBN: 9780170372855

3

Activity B

1 Write the word(s) that match each definition or description in the table below. Select from this list: *genetic, taxon, proboscis, adaptation, epidermis, transpiration, niche, gas exchange, organism, biochemical.*

a	Any feature that increases the chance of survival	
b	Chemical processes in living things	
c	The specialised 'occupation' of an animal or a plant	
d	A group of similar animal or plant species	
e	A living thing	
f	The evaporation of water from plant surfaces	
g	Diffusion of gases into and out of living things	
h	Outer skin	
i	Information inherited via DNA	
j	Tubular device for feeding or breathing	

2 Ruru (morepork owls), shown on the cover of this book, are nocturnal hunters of mice, large insects and small birds. They capture and kill prey with their sharp clawed feet. Their wing feathers have a downy covering that makes flight almost silent.

a Describe one physical adaptation of ruru beak shape.

b Explain one adaptive advantage of their beak shape.

c Explain one adaptive advantage of silent flight in owls, in relation to nocturnal behaviour.

d Forward-facing eyes make it possible for animals to judge distances accurately. Explain one adaptive advantage of this to ruru.

ISBN: 9780170372855

e Describe one other adaptive feature of ruru eyes.

f Explain one adaptive advantage of the feature you've just described.

Choosing a topic

Achievement Standard 91155 allows for a wide range of choices.

First, it is possible to choose between the following eight life processes:

internal transport; gas exchange; transpiration; nutrition; excretion; reproduction; support and movement; sensitivity and co-ordination.

Second, it is possible to choose what to compare; option A or B.

A Compare and contrast **one life process** across **three different groups** (taxa) of animals, or three different plant taxa. Example: support and movement in grazing animals (e.g. horses), primates (apes, etc), and big cats (tigers, etc)

or

B Connect **two life processes** within **one group** of animals (or plants). Example: how the gas exchange system and the transport system are linked in mammals.

This flexibility means that hundreds of different combinations are possible. Some schools may choose **B**, connecting nutrition and excretion in mammals; other schools may choose **A**, comparing and contrasting gas exchange or reproduction or movement in any three different kinds of mammal. Or reptiles. Or insects. Whatever is selected, each school will provide appropriate resources. Internal assessment could be through individual assignments, or by tests and exams.

This workbook does not attempt to cover all these many different possibilities. The next unit deals only with gas exchange. It also takes option **A**, with information on gas exchange systems in three groups: insects, fish, mammals. Further details on assessment requirements are given at the end of Unit 2.

Unit 2 | Gas exchange

With a few exceptions, most living things need O_2 for cell respiration and must get rid of the CO_2 produced by respiration. In addition, green plants need CO_2 for photosynthesis and must get rid of the O_2 it produces. In both situations, two gases are 'exchanged'. O_2/CO_2 **gas exchange** occurs in almost all plants and animals. Most animals have specialised gas exchange organs: **lungs** (on land), **gills** (in water).

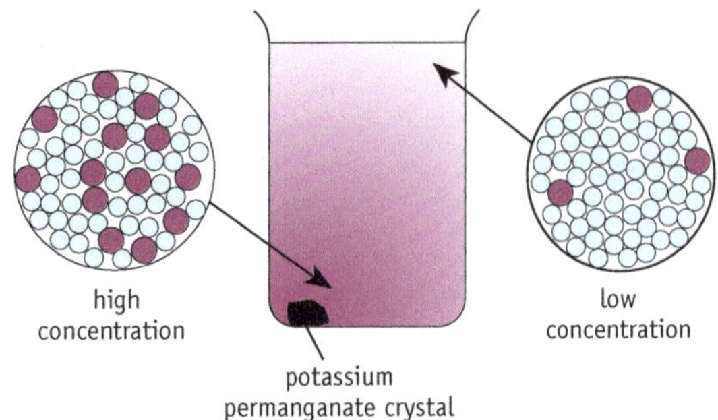

Fig. 2.3.1 Diffusion of a coloured substance in water.

3

Four important definitions, not to be mixed up

Gas exchange: the two-way diffusion of O_2 and CO_2 across a surface.

Diffusion: the movement of particles (atoms, ions, molecules) of any substance from where they are more concentrated to where they are less concentrated.

Cell respiration: biochemical energy-releasing processes that occur in most cells.

Breathing: body actions that move air in and out of the lungs/trachea; or water over the gills, in the case of fishes.

Problems with gas exchange

- Except over very short distances, diffusion is slow in air and far slower in solution.
- In land animals, O_2 must first dissolve in water before it can diffuse into the body.
- In land animals, a big thin moist gas-exchange surface loses a lot of water by evaporation.
- Water animals get their oxygen from dissolved O_2, not from the O in H_2O. Water contains very little dissolved O_2; at most 1 mL O_2 dissolves in each 100 mL of cold water. Warm water has even less. Air has 20% O_2.

Many different adaptations help increase the overall rate of gas exchange. The five general features listed below are shared by all vertebrate and many non-vertebrate animals, but details vary greatly.

Special feature of gas exchange surfaces in vertebrate animals	How it works
1 A specialised organ with its surface elaborately folded, increasing the total area.	A big surface area means larger amounts of gas can diffuse.
2 The surface is kept moist in land animals. In fish it is already moist.	Diffusion is faster through a liquid than it is through dry skin.
3 The surface is very thin, often a layer no more than one cell thick.	Shorter travel distance means a steeper concentration gradient.
4 Rich blood supply that continually removes O_2 and supplies CO_2.	Together, **4** and **5** help maintain a big difference in concentration across the two sides of the membrane. A 'steep concentration gradient' causes diffusion to be faster.
5 Breathing movements bring 'fresh' air to the gas exchange surface (or water, in the case of fish).	

ISBN: 9780170372855

Activity A

Complete sentences **1–4**.

1 Gas exchange surfaces tend to have a big surface area because ______________________

____

2 Gas exchange surfaces need to be kept moist because ______________________

____

3 Gas exchange surfaces like lungs tend to lose water rapidly because ______________________

____

4 Diffusion can be defined as ______________________

3

Comparing gas exchange in fish, mammal, insect

Activities B, C and D can be used as a guide to help compare these three different groups, but they provide only part of what you would need for your own portfolio on the subject. They deal with general adaptations; obviously there are different details in different species. Remember that to explain any particular adaptation involves dealing with two kinds of question: **how** (mechanism) and **why** (function/adaptive advantage).

Fish gill blood supply and water flow

This example shows the amount of detail needed at each grade. To achieve E, it's necessary to include A and M answer material as well.

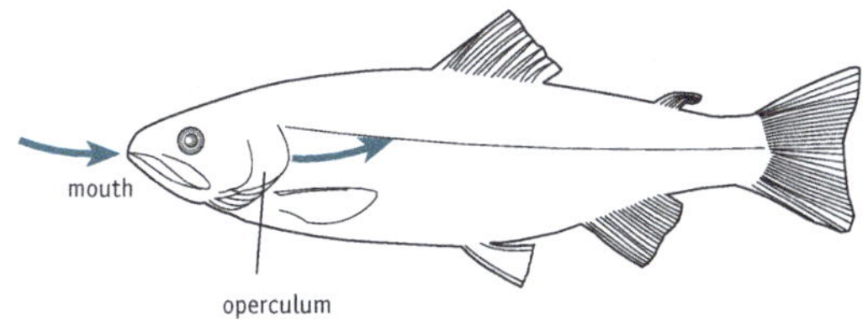

Fig. 2.3.2 Trout showing inhalant and exhalant water flow, resulting from breathing movements of the gill covers.

Achievement. Fish gills have lamellae and a rich blood supply. The lamellae give a big surface area to speed up O_2/CO_2 diffusion. Blood takes O_2 to the rest of the body.

Merit. Water has little dissolved O_2. One-way flow of water across the gills helps keep a steep concentration gradient of O_2, thus ensuring faster diffusion.

Excellence. Water is dense compared with air, which means that fish need to expend a lot of energy to move fast. Fish have a counter-current system with blood and water flowing in opposite directions, which helps ensure maximum oxygen absorption. On the other hand, fish don't need to do work against gravity so moving at moderate speed is not very energy-expensive. Limitation of structure: gills can only work in water because in air the lamellae collapse and stick together, which reduces their surface area and gas-exchange ability.

ISBN: 9780170372855

3

Activity B | Fish

Using trout gill drawings A–F, identify as many features as you can that help make exchange as efficient as possible. Draw a line from each feature; explain how it works; state its exact function. Refer to the five features in the table in the section 'Problems with gas exchange' (page 38).

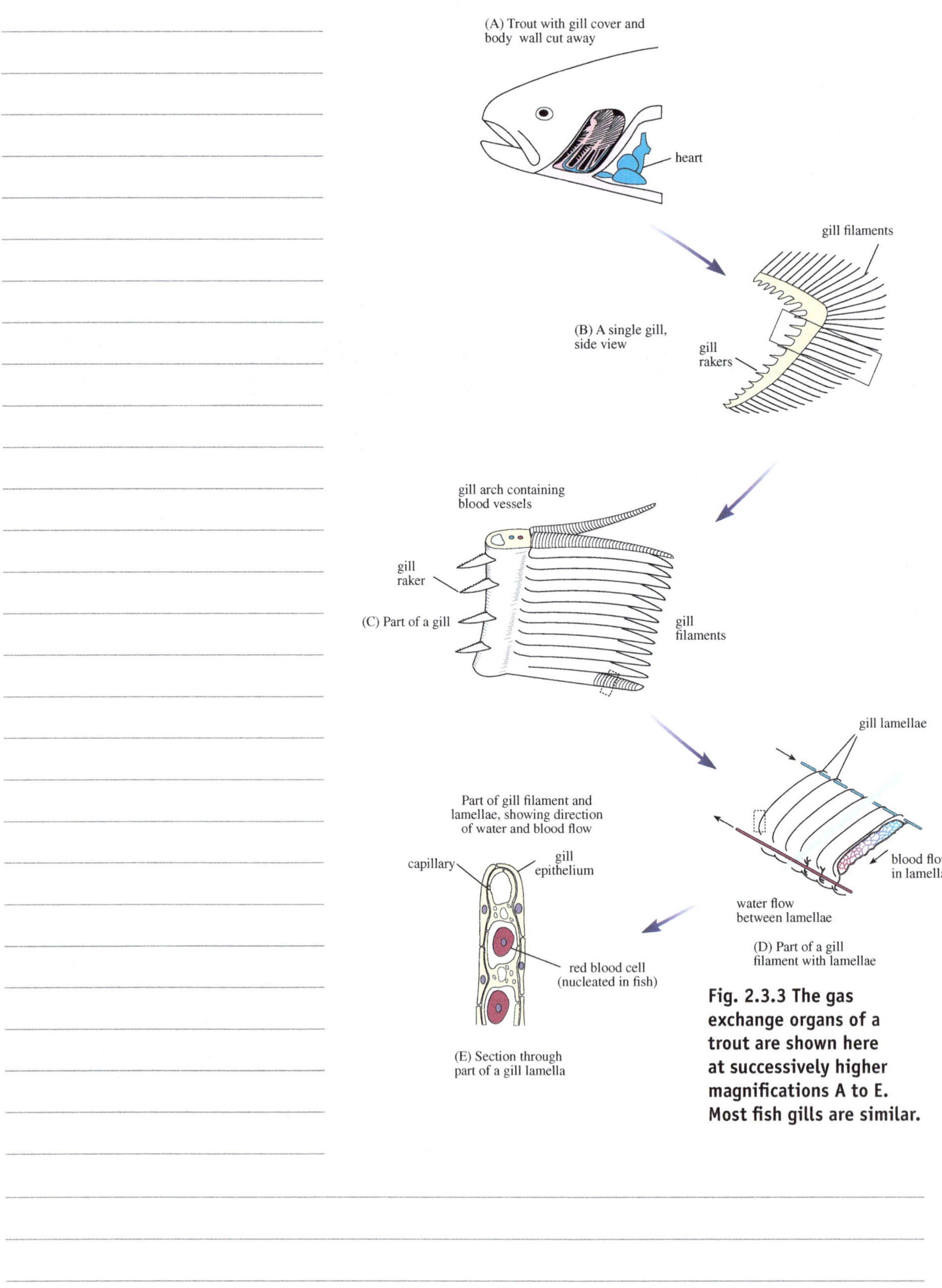

Fig. 2.3.3 The gas exchange organs of a trout are shown here at successively higher magnifications A to E. Most fish gills are similar.

ISBN: 9780170372855

Activity C | Mammals

Using human lung drawings 2.3.4, 2.3.5 and 2.3.6 plus the microscope photo (page 42), identify all visible features that help make exchange as efficient as possible. Draw a line from each feature; explain how it works; state its exact function. Refer to the five features in the table in the section 'Problems with gas exchange'.

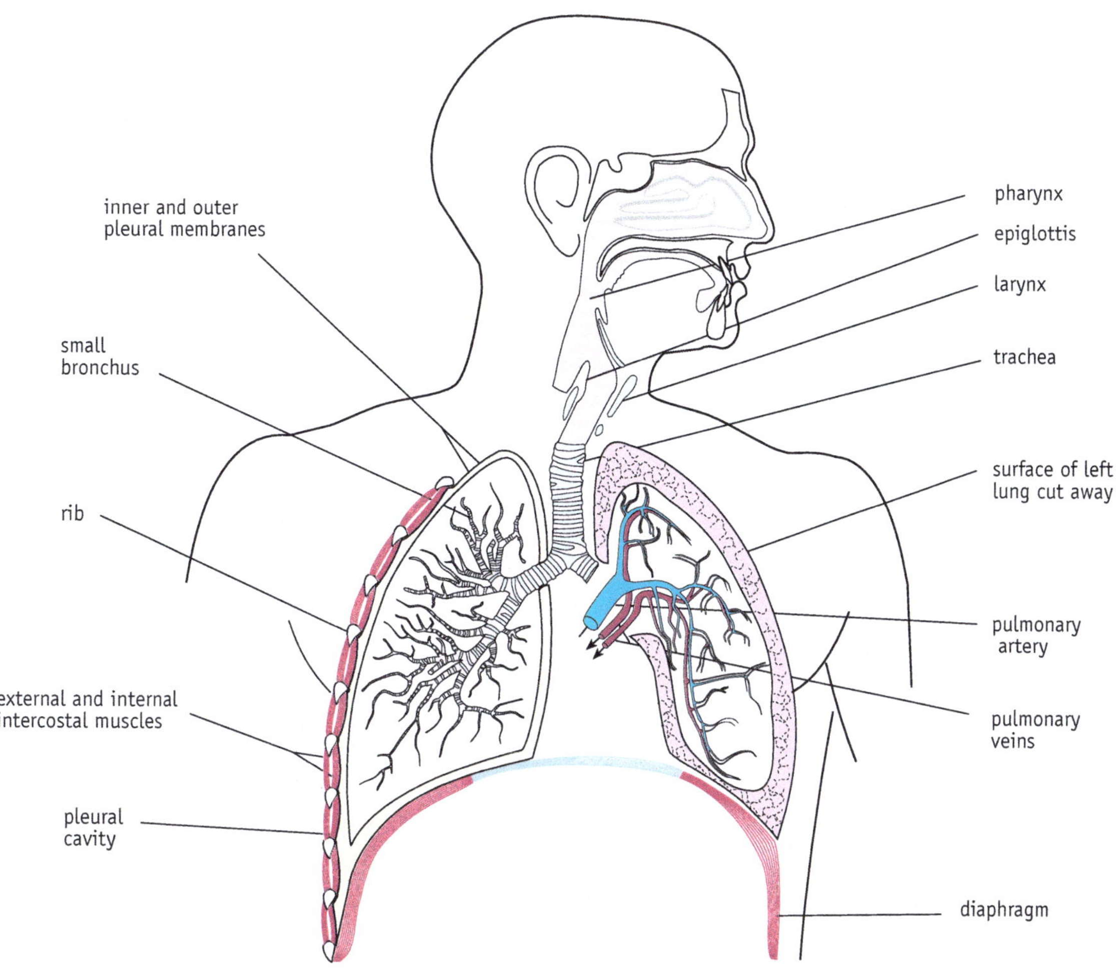

Fig. 2.3.4 Human lungs, also showing some surrounding structures and part of the blood supply. Most mammal lungs are similar.

E

The arrangement of trachea and bronchi means that 'tidal' air goes in and out the same tubes. Having residual air in these air passages causes poor air supply to the alveoli, which receive air low in oxygen. This is a definite limitation of the gas exchange system in mammals. The arrangement in birds is much more efficient, as it supplies continuous one-way air flow through the lungs. Result: good oxygen supply, so bird lungs are small compared with mammal lungs.

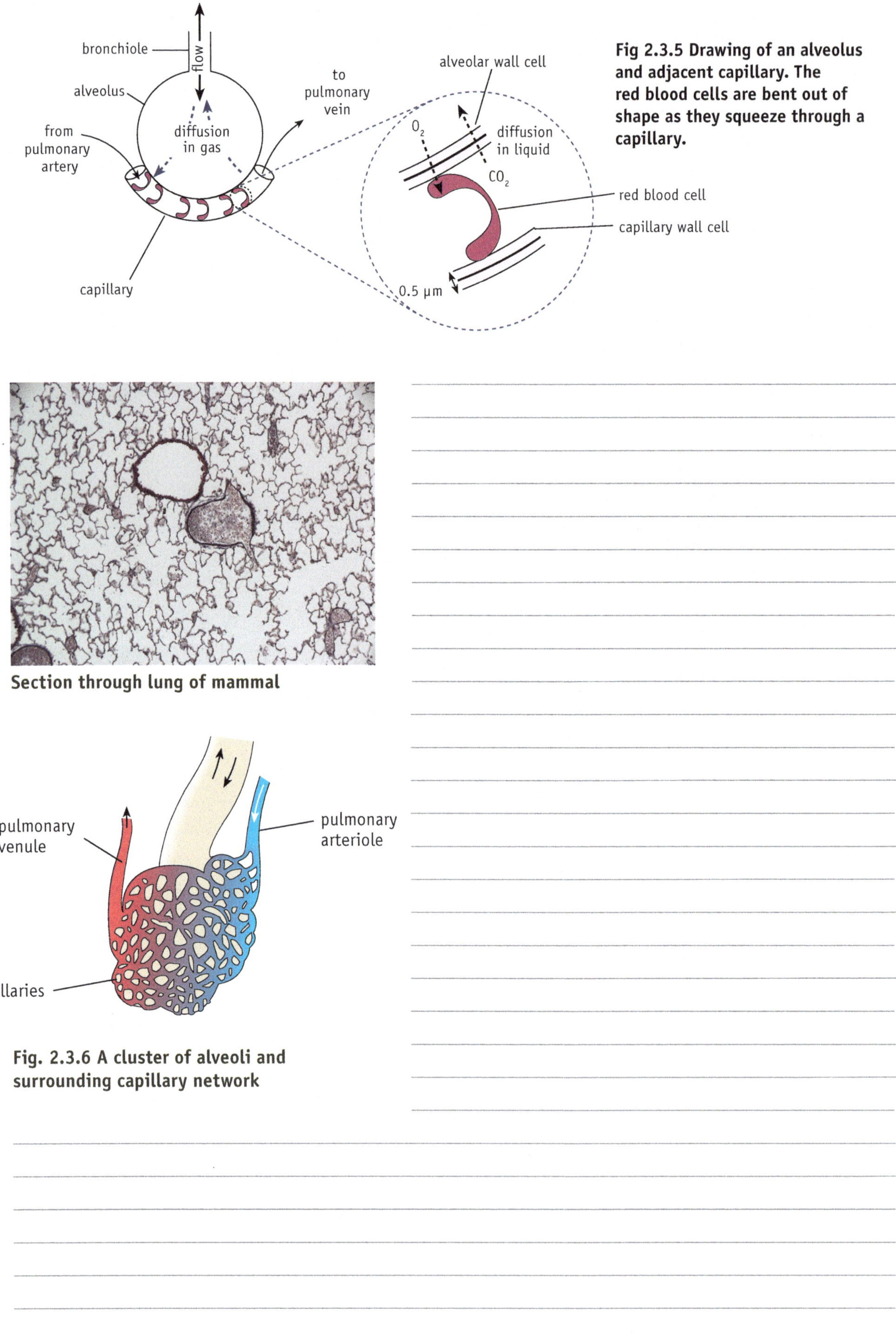

Fig 2.3.5 Drawing of an alveolus and adjacent capillary. The red blood cells are bent out of shape as they squeeze through a capillary.

Section through lung of mammal

Fig. 2.3.6 A cluster of alveoli and surrounding capillary network

To earn Excellence in this standard, one requirement is that you need to explain (where relevant) one limitation involving **structure**, plus one limitation imposed by the **environment**.

ISBN: 9780170372855

Activity D | Insects

Insect blood is not used to carry O_2 and CO_2. Instead, these gases are transported by tracheae, a system of branching air-filled tubes that reach throughout the body. Gas exchange happens only at the very ends. These tubes open at the sides of the body at spiracles, holes that can be closed when needed to reduce water loss. Pulsing movements of the body help pump air in and out of air sacs and the whole system.

Using insect drawings 2.3.7 and 2.3.8, identify as many features as you can that help make exchange efficient. Draw a line from each feature; explain how it works; state its exact function. Refer to the five features in the table in the section 'Problems with gas exchange'.

E

Trachea endings (tracheoles) are in contact with almost every cell in the insect body, which means that the endings are extremely thin and narrow. Oxygen diffuses inwards along these tracheoles, and CO_2 outwards. Limitation 1: this two-way diffusion is effective over a short distance but less effective over long distances. This probably puts a limit on the upper size of insects, and causes big insects to have slower metabolism than smaller ones. Limitation 2: having rings of chitin to support the trachea means that gas exchange cannot happen here.

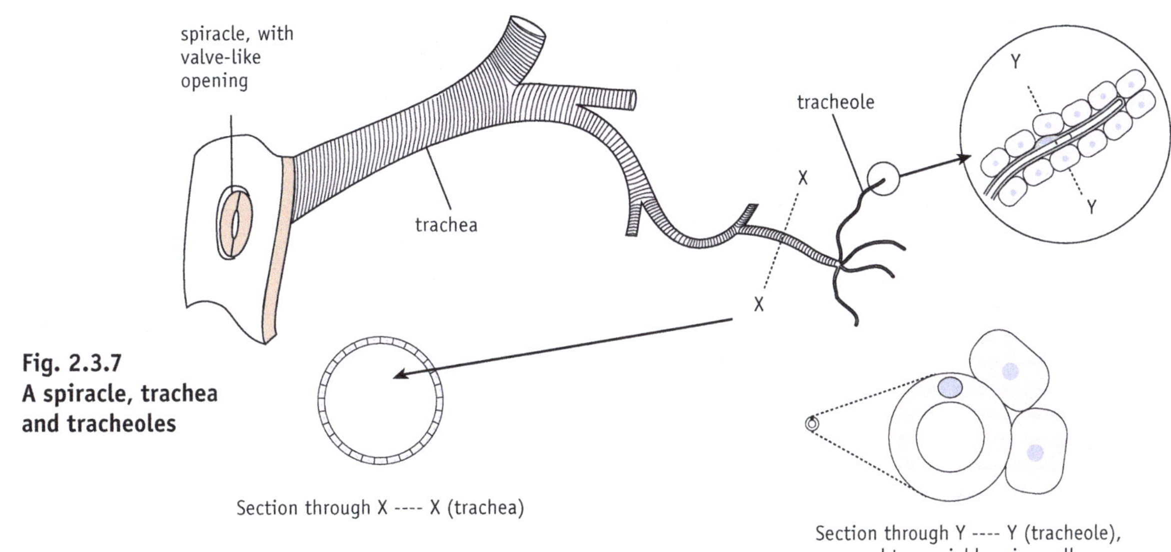

Fig. 2.3.7 A spiracle, trachea and tracheoles

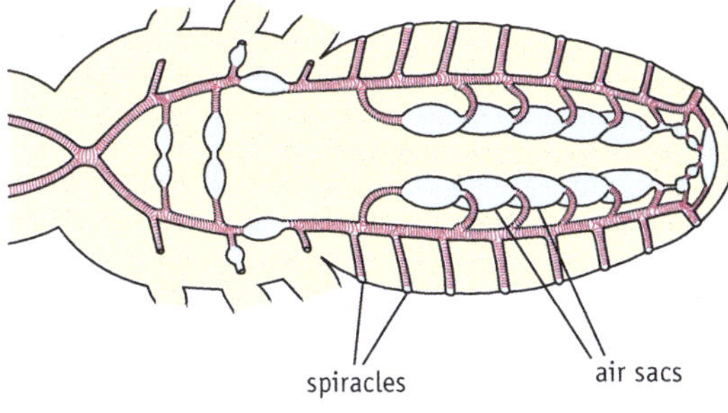

Fig. 2.3.8 Air sacs and main tracheae of a weta

3

3

Activity E

1 Write the word that matches each definition or description in the table below. Select from this list: *capillary, pulmonary, tracheole, diffusion, filaments, epiglottis, respiration, diaphragm, spiracle, larynx, pleural, rakers.*

a	These openings in insects lead to tracheae	
b	Small branch at the end of trachea	
c	Smallest blood vessels	
d	Major muscle that assists breathing movements	
e	Structures on fish, preventing food getting onto the gills	
f	Small gill structures that increase surface area	
g	Lubricating membranes surrounding the lungs	
h	Prevents food going down the trachea	
i	Blood vessels associated with the lungs	
j	Also known as the voice box	
k	Movement of particles to regions of lower concentration	
l	Biochemical energy-releasing processes	

2 The insect gas exchange system works very well for small insects, but may be a **limitation** for larger ones. This could be one reason why big insects tend to be slow-moving, and none is heavier than about 50 grams. Suggest what particular features of a tracheal system could make it inefficient for larger insects. Explain the mechanism of this limiting feature.

ISBN: 9780170372855

Biology 2.3 Plant and animal adaptations

AS 91155 Demonstrate understanding of adaptation of plants or animals to their way of life
Internally assessed, 3 credits

Achievement	Achievement with Merit	Achievement with Excellence
Demonstrate understanding of adaptation of plants or animals to their way of life.	Demonstrate in-depth understanding of adaptation of plants or animals to their way of life.	Demonstrate comprehensive understanding of adaptation of plants or animals to their way of life.

Achievement
'Demonstrate understanding ...' involves describing the adaptations and identifying the aspects of the adaptations that enable each organism to carry out its life process(es) in order to survive in its habitat.

Achievement with Merit
'Demonstrate in-depth understanding ...' involves providing a biological reason that explains how or why the adaptations enable each organism to carry out its life process(es) in order to survive in its habitat.

Achievement with Excellence
'Demonstrate comprehensive understanding ...' involves linking several biological ideas. The linking may involve justifying, evaluating, comparing and contrasting, or analysing, and must include consideration of the two points from below appropriate to the chosen context.

In the context of 'understanding of adaptation' related to one life process over three taxonomic or functional groups of multi-cellular plants or animals:

- comparing diversity of adaptation in response to the same demand across different taxonomic or functional groups
- limitations and advantages involved in each feature within each organism.

In the context of 'understanding of adaptation' across two related life processes within one taxonomic or functional group:

- connections between two life processes within each organism which enhance the effectiveness of both processes
- limitations and advantages involved in each feature within each organism.

Notes

1 'Understanding of adaptation' is demonstrated in relation to one life process over three taxonomic or functional groups of multi-cellular plants or animals, or across two related life processes within one taxonomic or functional group.

2 'Adaptation' involves the range of ways in which organisms have developed strategies to carry out the life processes. An adaptation refers to a feature and its function as it enables an organism to carry out a life process and thus occupy a specific ecological niche. It may include structural, behavioural, or physiological features of an organism. An adaptation provides an advantage for the organism in its specific habitat and ecological niche.

3 'Way of life' encompasses the ways in which an organism carries out all its life processes. It includes:
- relationships with other organisms — competition, predation, parasitism, mutualism
- reproductive strategies
- adaptations to the physical habitat.

4 'Life processes' are selected from:
- internal transport
- nutrition
- sensitivity and co-ordination
- gas exchange
- excretion
- reproduction.
- transpiration
- support and movement

Cell biology

NCEA Achievement Standard 91156

Demonstrate understanding of life processes at the cellular level

2.4

Externally assessed, 4 credits.

Unit 1 | Inside cells

4

All living things have cells, with around 10^{13} in your own body. At the microscope end of body size range, many kinds of organisms are one-celled.

Different cells have widely different shapes and specialised functions, which means there is no such thing as a 'typical' cell. However, at high magnifications we can see internal structures known as **organelles**, and these are much the same across almost all living things, both plant and animal. (Bacteria do not have organelles. Viruses are not even cells, and are not considered to be living things.)

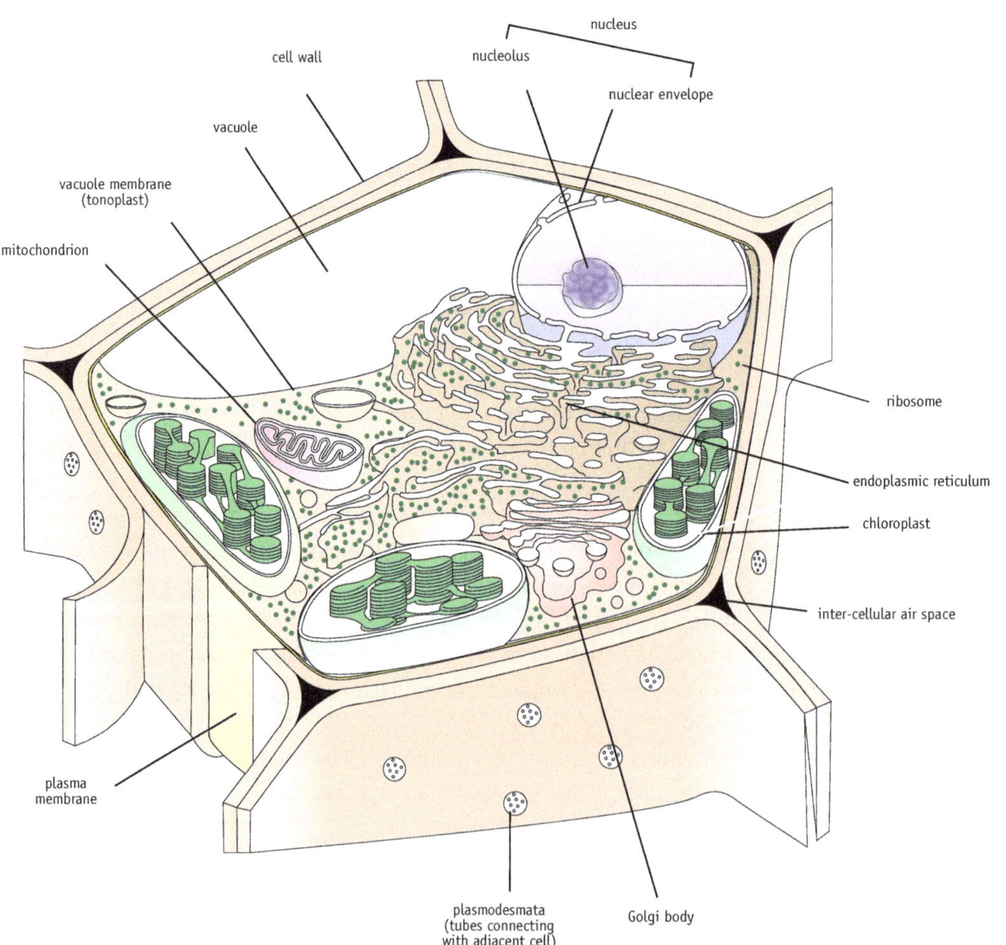

Fig. 2.4.1 A plant cell, enlarged here about 2000 times. The drawing shows some of its organelles, together with the cell walls of five neighbours. (You will not be expected to produce a drawing showing this amount of detail.)

ISBN: 9780170372855

Activity A

1 Complete this table. For **h–l**, read ahead to get the information you need.

	Organelle or cell part	Function or main activity: brief summary
a	Cell wall (in plants only)	To protect and support the cell
b	Vacuole (large in plants only)	Fluid pressure supports the cell; also sugar storage
c	Plasmodesmata (in plants only)	
d	Chloroplasts (in plants only)	
e	Mitochondria	
f	Plasma membrane	
g	Ribosomes	
h	Nucleus	
i	Nucleolus	
j	Endoplasmic reticulum	
k	Golgi complex	
l	Centriole (in animal cells)	

2 Plant/animal similarities and differences. In each of the empty boxes, write 'Yes' or 'In a few' or 'Never'.

Structure/organelle	In plant cells?	In animal cells?
Nucleus		
Mitochondria		
Ribosomes		
Endoplasmic reticulum		
Centriole		
Golgi complex		
Plasma membrane(s)		
Chloroplast		
Cell wall		
Large vacuole		

ISBN: 9780170372855

Plasma membrane

This forms the outermost living layer of the cell — but inside the cell wall, if any. The same type of membrane surrounds and forms all organelles. A plasma membrane (previous name 'cell membrane') consists of a double layer of lipid (fat) molecules, and also proteins.

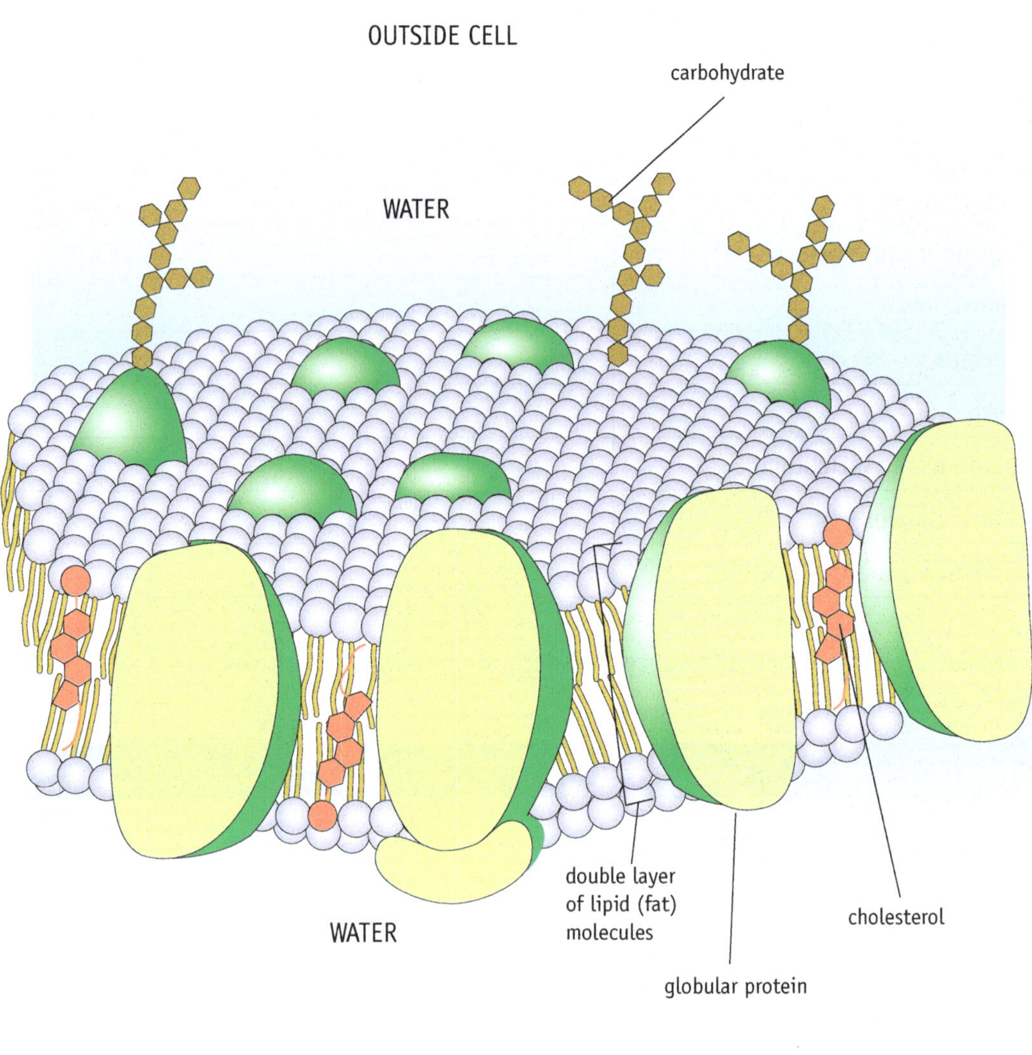

Fig. 2.4.2 Structure of a plasma membrane. Special protein molecules extend into and sometimes through this membrane, and have an important role in the uptake of chemicals.

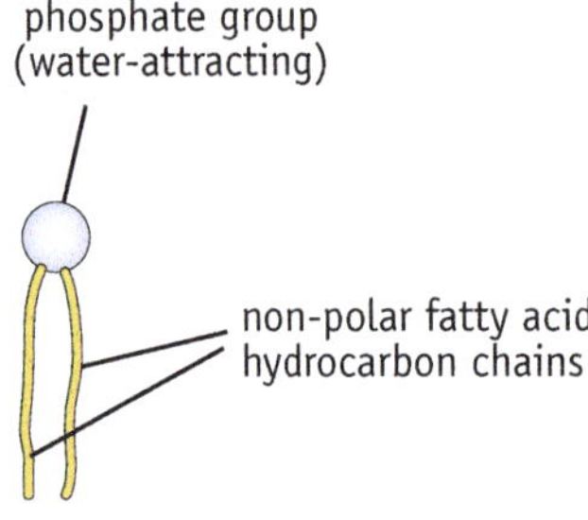

Fig. 2.4.3 A single phospholipid molecule in the plasma membrane. The 'head' contains a phosphate group and is attracted to water molecules. The fat-soluble 'tails' form the inner part of the membrane.

4

ISBN: 9780170372855

Plasma membranes regulate the rate at which substances enter and leave each cell. Plasma membranes are partly permeable (semi-permeable), and small molecules such as O_2 can diffuse through easily. Larger molecules like sugars are transported across the plasma membrane by special proteins in the membrane. Unit 2 has information on different transport mechanisms. The plasma membrane is also responsible for receiving signals from other cells.

Nucleus

A nucleus holds genetic information in the form of DNA, which is located in pairs of threadlike **chromosomes**. The nuclear envelope (modern name for nuclear membrane) has tiny pores through which substances pass between nucleus and cytoplasm. Each nucleus also contains a dark **nucleolus**, in which ribosomal RNA is formed.

Endoplasmic reticulum (ER)

ER is a three-dimensional network of membrane-bound spaces — the spaces themselves being known as cisternae (singular: cisterna). There are two types of ER:

- **rough endoplasmic reticulum** (RER) has large numbers of **ribosomes.** These tiny particles are the sites of protein synthesis. Cells active in making proteins are rich in RER.
- **smooth endoplasmic reticulum** (SER) lacks ribosomes. SER is involved in the synthesis of lipids, such as steroid sex hormones.

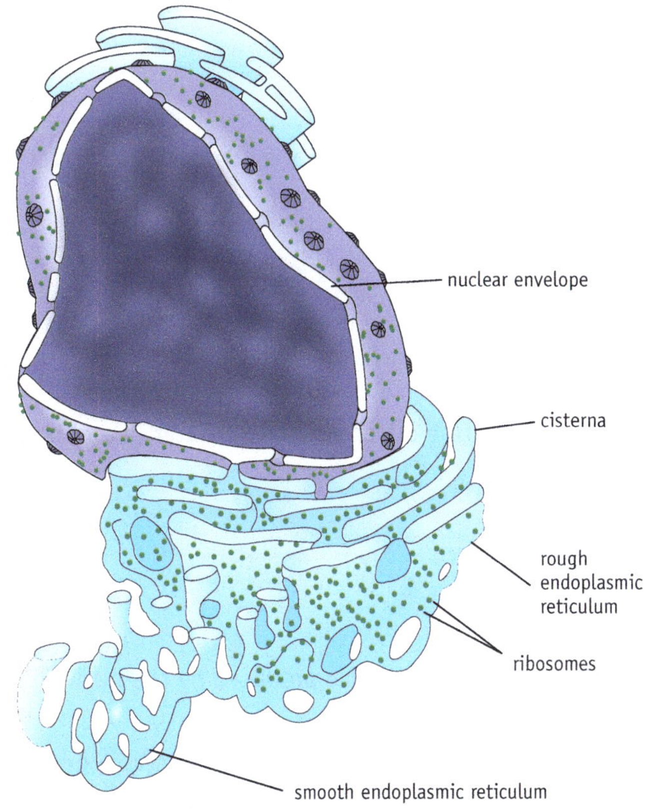

Fig. 2.4.4 The nuclear envelope is continuous with the endoplasmic reticulum. The shape of the ER provides a large surface area for ribosomes, and the cisternae provide a transport function within the cell.

Golgi complex

After a protein molecule is made in ribosomes it may undergo further treatment. Some molecules are labelled, enabling them to be directed to their correct destination. These 'treatments' happen in the Golgi complex (aka Golgi). Cells specialised for protein secretion may have several of these.

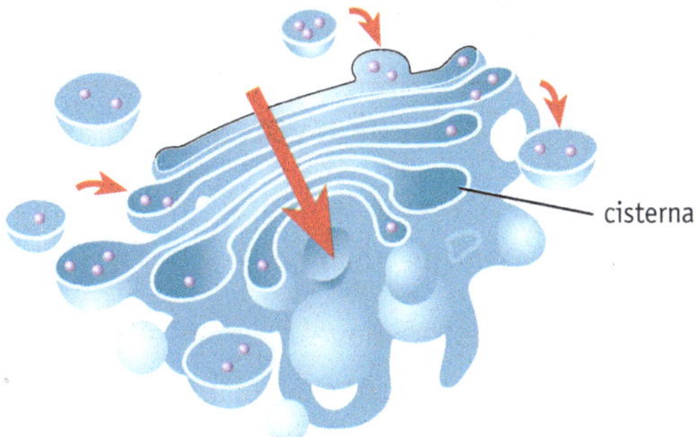

Fig. 2.4.5 A Golgi complex is similar to smooth ER. After completion of protein synthesis, little vesicles (packets) of protein are then transported to their destinations.

> The watery gel between and around all the organelles is known as **cytosol.**

4

ISBN: 9780170372855

Activity B

1 Name four cell organelles that are surrounded by membranes.

2 Explain what is meant by 'semi-permeable', in a cell context.

3 Name three examples of molecules that travel freely through plasma membranes.

4 a What do the letters 'ER' stand for, in a cell context?

b Describe differences in appearance and function between SER and RER.

5 Name one substance produced by the nucleolus.

6 Describe and explain the functional link between a Golgi complex and ER.

7 State where chromosomes are located.

Prokaryotes and eukaryotes

Prokaryotes are simple living things; their cells do not have organelles. Their single chromosome forms a loop and is not enclosed by a nuclear envelope. Example: bacteria.

Eukaryote cells do have organelles and a nucleus. Their ribbon-like chromosomes are open-ended. Most living things (not bacteria) are eukaryotes.

 ISBN: 9780170372855

Mitochondria

Each mitochondrion is a 'powerhouse', a site of respiration in which most of the cell's ATP is produced. ATP is the direct source of energy for almost all cell activities: muscular action, active transport, protein synthesis, etc.

Mitochondrial numbers depend on each cell's chemical activity. Heart muscle cells require a great deal of energy so have many mitochondria, while red blood cells have none.

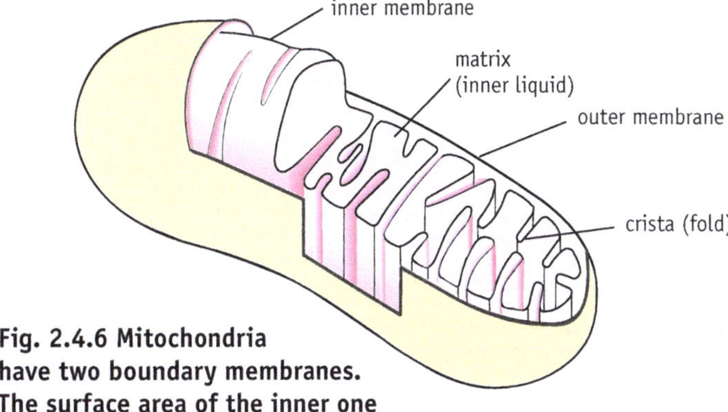

Fig. 2.4.6 Mitochondria have two boundary membranes. The surface area of the inner one is increased by folds called cristae. The enzymes and chemical processes of respiration occur on these surfaces.

Like chloroplasts, mitochondria contain their own DNA, and are thought to have originated from bacteria in the distant past. Mitochondria are about 1 μm wide and 2–3 μm long. Their long thin shapes help increase surface area, which speeds the uptake of oxygen and fuel.

The function of mitochondria is to make molecules of ATP. The process starts outside mitochondria with **glycolysis**, and continues inside with **respiration**. Together these two processes can be greatly simplified as:

glucose + O_2 → CO_2 + H_2O + energy used to 'recharge' ATP + waste heat energy

ATP functions as a temporary 'energy holder'. Its recharge-and-use cycle is simplified in Fig. 2.4.8, with 'P' representing a phosphate group.

4

E

More on respiration

The breakdown of glucose begins outside each mitochondrion, starting with the process of **glycolysis**. This produces 3-carbon pyruvate ions that then enter the mitochondrion as fuel. Oxygen-using (aerobic) **respiration** happens inside the mitochondria, producing 28 of the 30 ATP molecules resulting from the breakdown of each molecule of glucose.

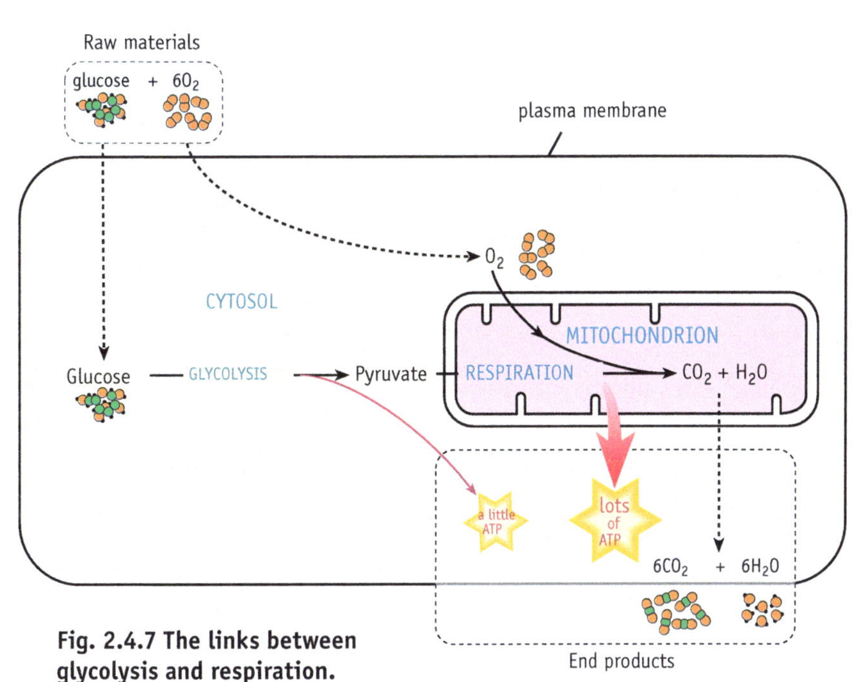

Fig. 2.4.7 The links between glycolysis and respiration.

This includes the **Krebs cycle** and the oxidative processes of hydrogen transfer.

ISBN: 9780170372855

For Achievement Standard 91156 you are not required to know details of glycolysis, fermentation (aka anaerobic metabolism), the Krebs cycle, nor any biochemical details of respiration. It is, however, helpful to have an understanding of the roles of ATP (adenosine triphosphate) and ADP (adenosine diphosphate).

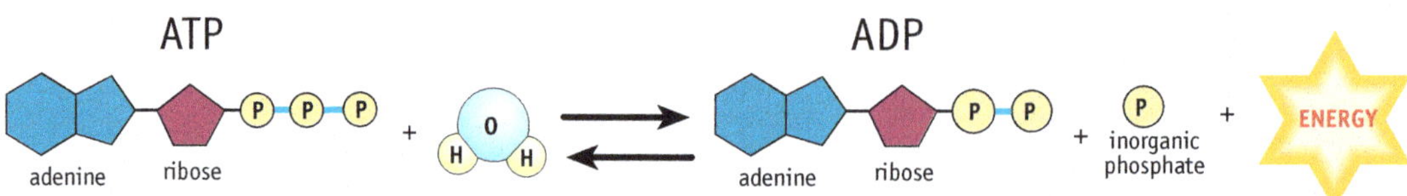

Fig. 2.4.8 Your body's small amount of ATP is continuously being recharged from ADP + phosphate. Energy for 'recharge' is provided by glycolysis and respiration. ATP's energy is then available for any processes inside the cell.

Chloroplasts

4

Chloroplasts are solar-powered sugar factories, the place where photosynthesis happens. They are abundant in places like leaf cells, and are particularly numerous in the outer zones of cells where they can get more CO_2.

Overall, the process of cell photosynthesis uses simple inorganic molecules and light energy, and produces sugar molecules that act as a temporary store of chemical energy. The process can be summed up as:

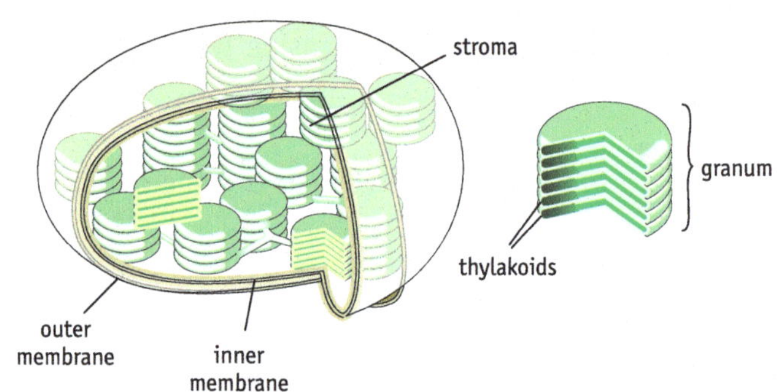

Fig. 2.4.9 Internal structure of a chloroplast. Light-dependent reactions of photosynthesis occur in the thylakoids (which contain chlorophyll), and light-independent reactions occur in the stroma.

CO_2 (from air) + H_2O (from soil) + **light** → **sugars** + O_2

Hands-on experiment: photosynthesis

Biology 2.1, Unit 2, Activity B describes a method of measuring the rate of photosynthesis. You could use this setup for your own investigations. Two questions that could be tested by experiment:

1. How much does photosynthesis rate increase when light intensity is increased?
2. How much does photosynthesis rate increase when the plant already has high light intensity, and extra CO_2 is supplied in the form of sodium carbonate solution?

Also use Biology 2.1, Unit 1 as a guide.

Lysosomes

Lysosomes are tiny packets of digestive enzymes, broken away from the Golgi complex. These enzymes are used to digest worn-out organelles such as mitochondria, and in some cases to digest and destroy invading bacteria.

ISBN: 9780170372855

Cilia and flagella

These are mobile hair-like extensions of the plasma membrane. In one-celled organisms they are used for swimming; in oysters they are used for filter feeding. Humans have cilia lining air passages and also the oviducts, where they help sweep each egg towards the uterus.

E

Word origins

Most organelle and cell names are based on Greek and Latin words. Knowing their origins can help. *Chloro*: green. *Chromo*: colour. *Soma*: body. *Flagellum*: whip. *Cilium*: eyelash. *Lysis*: destruction. *Mitos*: thread. *Khondrion*: small granule. *Endo*: inside. *Rete*: net. *Kytos (cytos)*: a container.

Activity C

1 Complete sentences **a–k** in fewer than 10 words each.

a The primary function of chloroplasts is to ____________________

b The primary function of mitochondria is to ____________________

c One similarity between the internal structure of chloroplasts and mitochondria is ____________________

d The difference between 'chloroplast' and 'chlorophyll' is ____________________

e A simple word equation for cell respiration is ____________________

f A simple word equation for photosynthesis is ____________________

g The main function of lysosomes is ____________________

h The energy-holder known as ATP is 'recharged' from ADP and phosphate, using energy obtained from ____________________

4

i A function of cilia and flagella in micro-organisms is to ______________________

____

j The main difference between prokaryote and eukaryote cells is ______________________

____

k The likely reason why mitochondria are long and thin is ______________________

____

2 Compare photosynthesis and cell respiration by writing words or symbols in the blank spaces.

	Energy source?	Gas used?	Gas made?	Organic molecules
Photosynthesis				made
Respiration				broken down

3 Suggest two kinds of cell that have the most mitochondria, and two that have the least. In each case, give a reason for your choice. Select from this list: *fat cells, heart muscle cells, bone cells, liver cells.*

a Two with most mitochondria:

____

____

b Two with least mitochondria:

4 Add words/symbols to this outline of a mitochondrion. Select from: *CO_2, O_2, pyruvate, ATP, ADP, phosphate.*

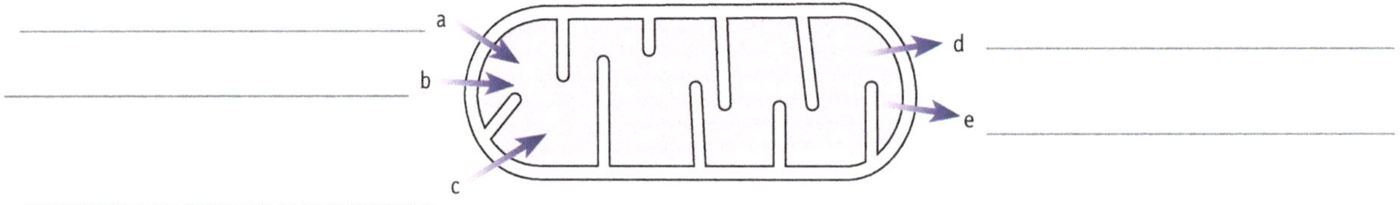

5 Add words/symbols to this outline of a chloroplast. Choose from: *CO_2, O_2, sugars, water, light.*

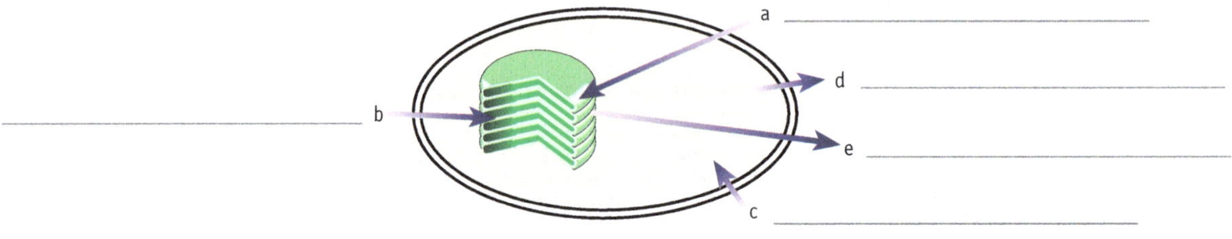

 ISBN: 9780170372855

6 Complete the following table.

Cell component	Description	Function
Cell wall		Support; prevents cell bursting in dilute solutions.
Plasma membrane	Double layer of phospholipids; on electron micrographs appears as two dark lines.	
Microvilli		Increase surface area of cells specialised for absorption.
granum, thylakoid, stoma Chloroplast	Plant cells only. Smooth outer membrane. Flattened sacs called thylakoids in piles called grana. Liquid stroma between grana.	
matrix, crista Mitochondrion		Produces ATP in respiration.
pore, nucleolus, nuclear envelope Nucleus	Surrounded by double-layered nuclear envelope, perforated by pores. Contains DNA in chromosomes.	
ribosome, cisterna Endoplasmic reticulum		Involved in distribution of substances in cytoplasm. Lipid synthesis (smooth ER).
Lysosome		Digestion of worn-out organelles. In some, white blood cells digest bacteria.
Ribosome	In electron micrographs, appear as black dots. Some are free in cytosol, most are attached to endoplasmic reticulum.	
Cilia	Hair-like extensions of plasma membrane, containing microtubules.	
Golgi body	Flattened membrane-bound sacs with small vesicles budding off the edges.	

ISBN: 9780170372855

7 For the mitochondrion, describe how the three raw materials enter, and where each comes from. Also outline what factors might affect the rate of transport of these three substances.

8 For the chloroplast diagram in question 5, describe how each of the two 'inputs' enters the organelle, and state where each one comes from. Also state three or more final 'destinations' of substances derived from the main organic product.

 ISBN: 9780170372855

Unit 2 | Size and scale

This page shows a range of sizes from humans down to molecules. Each measure shown on the left side is 10 times greater than the one below it, so the range here is 10^{11}.

It is not practical to measure cell and organelle sizes in millimetres or centimetres, so smaller units are used: µm and nm.

micrometre, symbol **µm**

1 mm = 1000 µm

nanometre, symbol **nm**

1 µm = 1000 nm

Nanometre units are used for objects such as viruses and molecules.

To give an idea of the size range of microscopic objects, the difference between micrometres and mm is 1000-fold, equivalent to the difference between mm and metres.

Biology 2.8 (page 204) gives details on different types of microscope. Light microscopes are used for objects down to about 1 µm size. Electron microscopes can show details in objects down to about 10 nm.

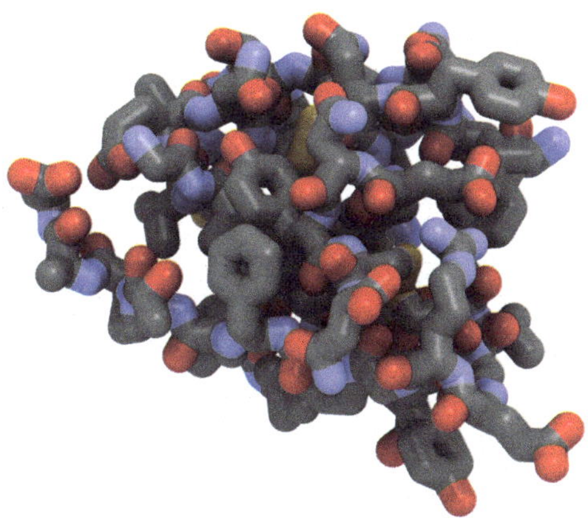

Insulin, a small protein molecule, is less than 1/1000 the diameter of a red blood cell.

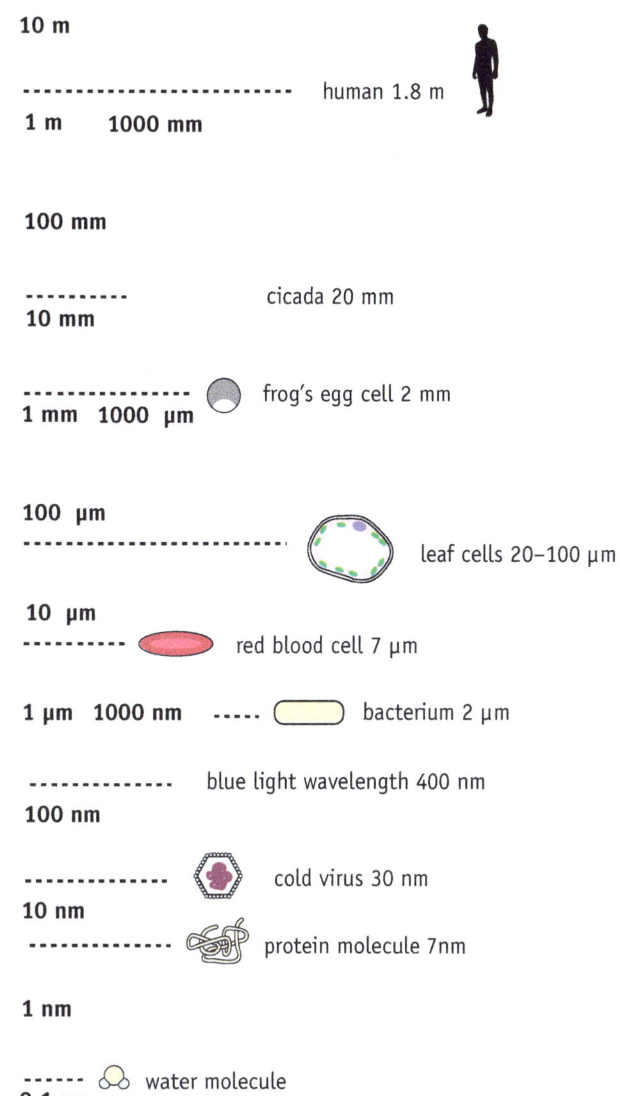

Fig. 2.4.10 A range of sizes encountered in biology. Notice the scale is logarithmic; each size unit differs 10-fold from the next.

4

ISBN: 9780170372855

4

Cell photo

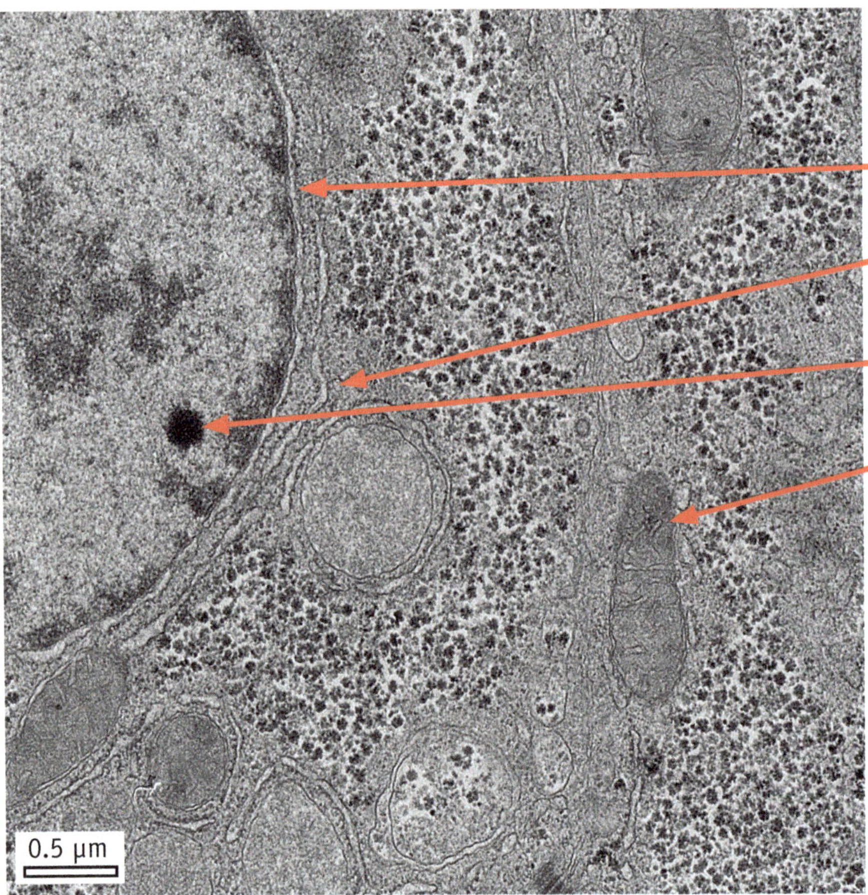

Photograph taken with an electron microscope, showing part of a liver cell. Part of the nucleus occupies the top left area. The many irregular dark patches outside the nucleus are glycogen, a carbohydrate stored by the liver.

Activity A

1 Identify structures and organelles **a–d** in the photograph above.

a ______________________ **b** ______________________

c ______________________ **d** ______________________

2 Using the scale shown, measure the length of organelle **d** in µm (micrometres, 1/1000 mm).

____

3 Measure the photograph length of **d** in mm: ____________ mm

4 Using your answers to questions **2** and **3**, calculate the enlargement in the photograph, compared with the original cell.

____

____

____

ISBN: 9780170372855

Activity B

The drawing shows an animal cell. Write in the names of parts **1** to **12**.

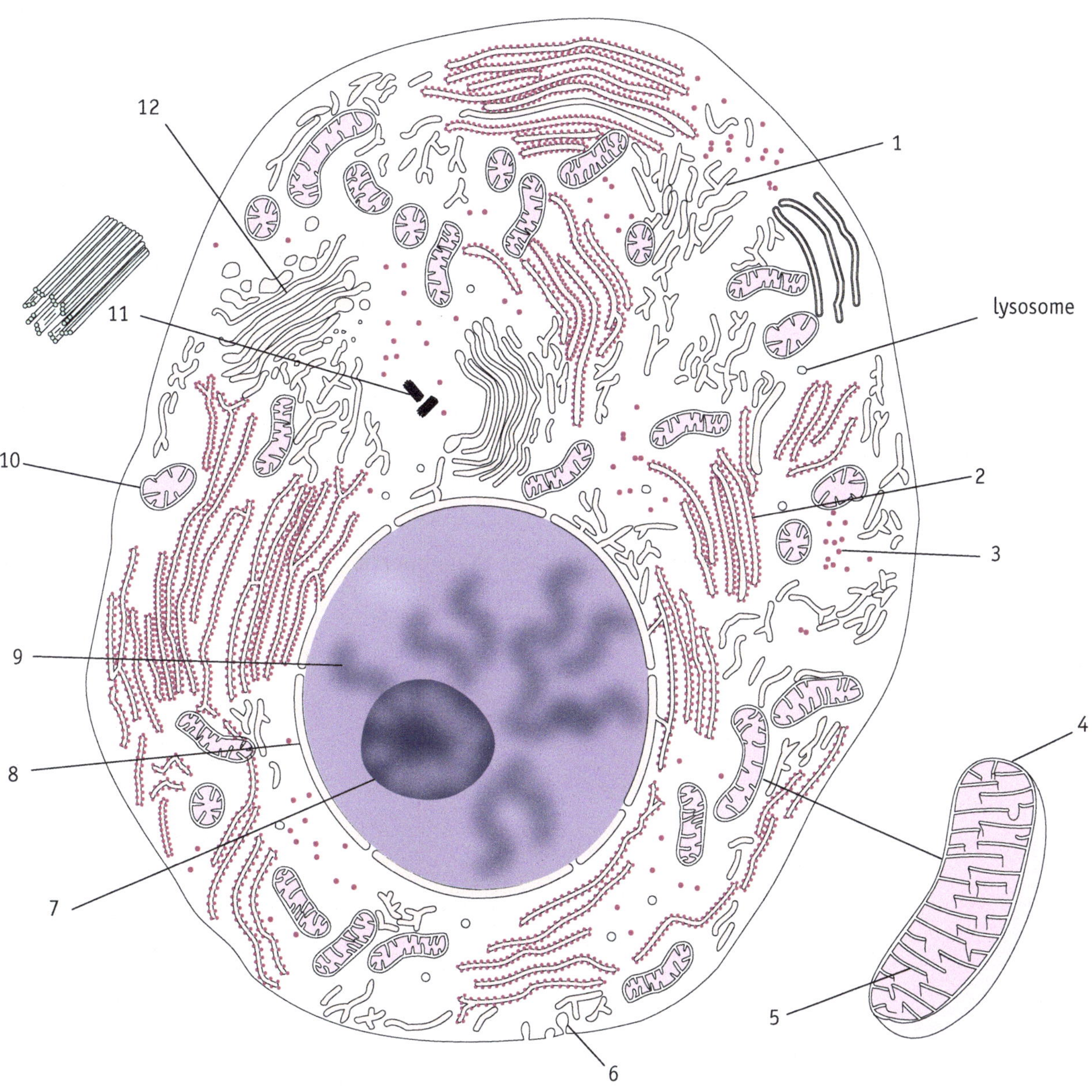

ISBN: 9780170372855

Activity C | Hands-on

This experiment uses cubes of agar jelly as models for cells. Refer to Biology 2.1 for guidelines on experiment design.

Question: Why are cells so small?

Aim: To test the hypothesis that if cells have a bigger surface area, then perhaps this increases the rate at which oxygen enters (diffuses in).

Method:

- Your teacher will provide each group with a slab of firm agar gel at least 3 cm x 3 cm x 5 cm. The gel includes a few drops of phenolphthalein indicator. (25 g agar dissolved per litre hot water; while still warm, stir in about 3 mL phenolphthalein and three drops dilute HCl. Pour the mixture into plastic tray so that the final depth of set gel is 3 cm. Store in fridge for up to a week.)
- From this slab, use a small kitchen knife to neatly cut one big piece with 3 cm sides, a medium piece with 2 cm sides, and a small piece with 1 cm sides.
- Pour about 150 mL 0.1 M NaOH solution into a 250 mL beaker.
- Place all three cubes in the same beaker simultaneously. Leave them for three minutes.
- Pour out all the NaOH solution, then immediately replace it with plain tap water, then rinse the cubes two more times. Make sure you don't pour them down the sink!
- Place all three cubes on flat white paper, then cut each through the middle, making a 3 mm-thick slice of each. For each slice, draw which zones are pink and which are not.

Results: Draw what you see. Also estimate (or calculate) the proportion of pink in each 'cell'.

Calculations: Complete this table.

	Small cube	Medium cube	Big cube
Length of cell	1 cm	2 cm	3 cm
Total surface area	6 x (1 x 1) = 6 cm^2	6 x (2 x 2) =	6 x (3 x 3) =
Total volume	1 x 1 x 1 = 1 cm^3		
SA:V ratio	6:1		

 ISBN: 9780170372855

Conclusions:

1 Make a general statement summarising how effectively diffusion of the NaOH supplied the interior of each cube 'cell'.

____

____

____

2 Complete this general statement about your calculation results: 'Small objects have a

______________ surface area-to-volume ratio, compared with ______________

objects of the same shape.'

3 Which cube had the biggest surface area? ______________ Which cube absorbed

the NaOH to the greatest extent? ______________

4 Do your results support the original hypothesis? Y/N ______

5 Make a general statement that links your observed results with calculated SA:V ratios.

____

____

____

____

____

____

____

Discussion:

Discuss your results and conclusions as they apply to actual cells, which are mostly much less than 0.1 mm diameter. In your discussion, refer to named substances and how each enters and exits cells.

____

____

____

____

____

____

____

____

____

ISBN: 9780170372855

Unit 3 | Transport and enzymes

Knowing the names and functions of cell organelles is a start, but it's more important to understand how they work.

All cells continuously need to take chemicals from their surroundings, and move other chemicals out. Five methods are described; all involve plasma membranes.

- Three **passive** methods of transport are: diffusion, osmosis, facilitated diffusion.
- Two **energy-using** methods are active transport and cytosis.

Diffusion

Diffusion is the movement of a substance from a region of higher concentration to a region of low concentration. The movement is caused by the kinetic energy of particles (molecules and ions). Example: oxygen being absorbed.

The rate of diffusion depends on several factors:

- Diffusion is thousands of times faster in a gas than in liquids.
- Diffusion is faster where there is a 'steep concentration gradient'; which means a big difference in concentration over a short distance.
- Diffusion is faster at higher temperatures.

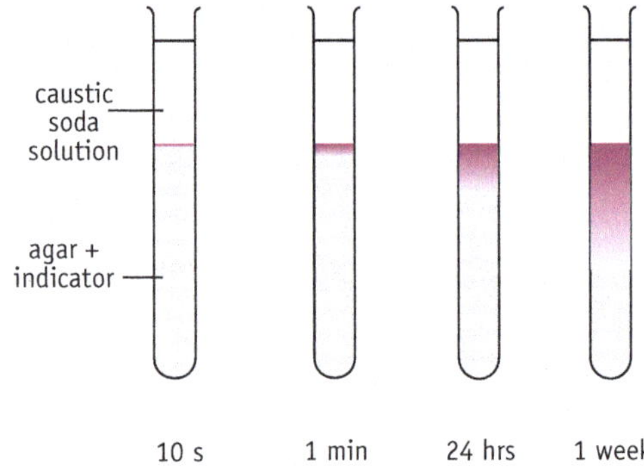

Fig. 2.4.11 Agar gel in the test tube has had a small amount of phenolphthalein indicator added. NaOH solution added above the gel results in slow diffusion and the spread of pink colour.

Osmosis

Osmosis is a special case of diffusion.

Definition: Osmosis is the movement of water through a semi-permeable membrane (spm), away from the side with higher water potential towards the side with lower water potential.

'Water potential' depends on the concentration of the water, and also on its pressure. Put simply: water moves from a 'watery' (dilute) solution into a more concentrated solution: D→C. Put more technically: water diffuses from a hypertonic solution to a hypotonic solution.

hypotonic: having a lower concentration of dissolved substances
hypertonic: having a higher concentration of dissolved substances
isotonic: having the same concentration of dissolved substances
turgid: describes any cell inflated with water pressure
flaccid: describes any cell deflated
semi-permeable membrane: one that allows water molecules to pass through, but not larger molecules such as sugar.

4

ISBN: 9780170372855

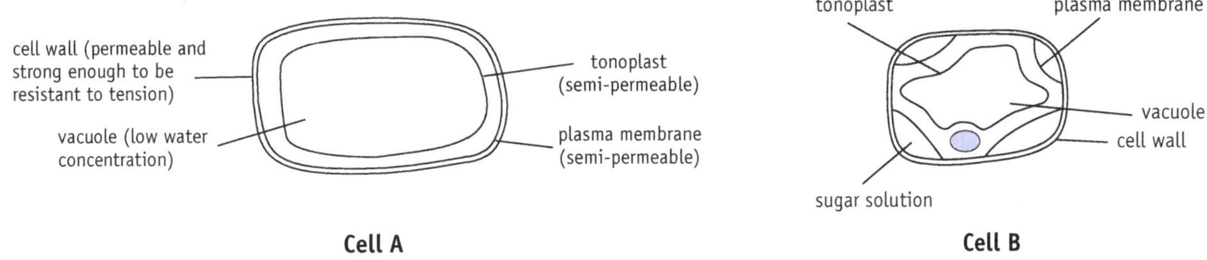

Fig. 2.4.12 Cell A has been placed in a more dilute solution (hypotonic to the cell). Result: the cell has absorbed water by osmosis and the vacuole is inflated (turgid). Cell B has been placed in a more concentrated sugar solution (hypertonic to the cell). Result: water has moved out of the cell; its vacuole and cell contents have shrunk. The cell is said to be plasmolysed, and the surrounding tissues will become flaccid.

Active transport

Active transport is the movement of molecules or ions across a membrane 'uphill' against a concentration gradient. To do this, the cell has to expend energy (from ATP). Example: the absorption of glucose by cells lining your intestine.

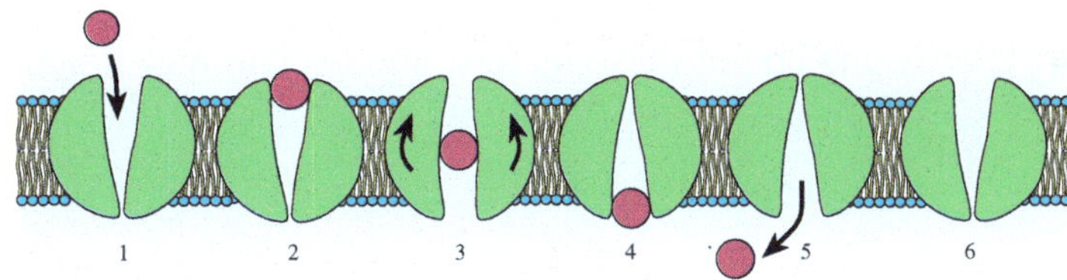

Fig. 2.4.13 A mechanism of active transport. This involves specialised proteins acting like 'pumps' in the cell membrane. Each particle (red) combines briefly with the membrane protein before being pulled across the membrane.

Facilitated diffusion

This process is similar to active transport in that it involves carrier proteins in a membrane. The difference: this is a 'downhill' process with particles moving away from regions of higher concentration — so the cell does not need to spend energy. In mammals, most cells absorb glucose by facilitated diffusion.

Cytosis

Cytosis involves bulk transport across a membrane in which the membrane is folded around the material being transported. **Pinocytosis** involves liquids. For larger particles and solids we call the process **phagocytosis** (*phagos* means 'to eat'). Example: white blood cells eating bacteria.

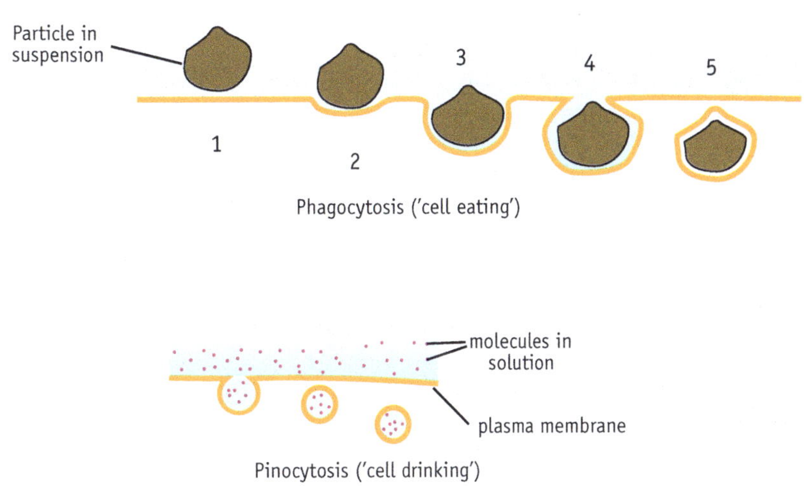

ISBN: 9780170372855

Activity A

1 List three passive methods of transport in cells. ______

2 Explain why they are described as 'passive'. ______

3 List three factors that speed up the rate of diffusion. ______

4 Choose any one of these factors, and explain how it speeds up diffusion. ______

5 Name two kinds of transport that require metabolic energy. ______

6 Explain the role of ATP in active transport. ______

7 Transport proteins are very specific. Explain what is meant by the underlined words. ______

8 Movement of molecules across a membrane is 'uphill' in some situations. Explain what is meant by 'uphill' in this sense.

9 For each of the following descriptions, identify the kind of transport mechanism to which it refers.

  a Oxygen moving into a cell. ______

  b A cell shrinking when placed in a hypertonic solution. ______

  c Involves protein carriers but does not need energy from metabolism. ______

ISBN: 9780170372855

d Movement of molecules across a membrane, using metabolic energy. ______________

e Movement of any substance from higher to lower concentrations. ______________

f White blood cell swallowing bacteria. ______________

10 This diagram represents a membrane surrounding a muscle cell.

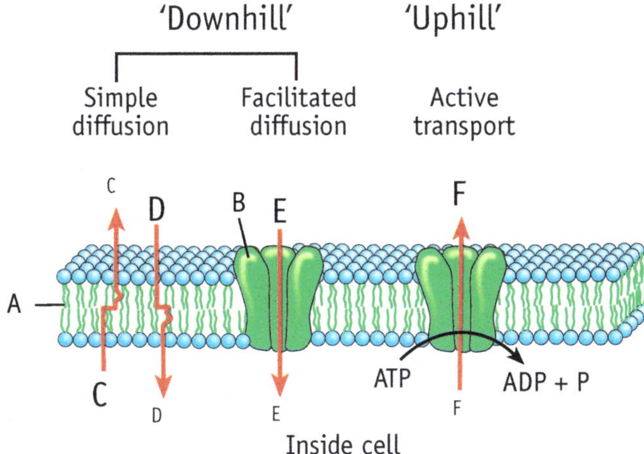

First, label parts A and B.

A ______________________________ B ______________________________

Next, against each of arrows, write the name of specific molecules that follow paths C–F through the membrane. Choose from this list: *O_2, protein, glucose, CO_2*.

Finally, draw in a section of membrane to show cytosis by extending the diagram a few centimetres to the right.

Hands-on experiment: osmosis

Biology 2.1, Unit 2, Activity C describes a method of investigating osmosis by measuring the tendency of water to move in and out of cells under different surrounding conditions. The results should give an indication of what solution concentrations are in osmotic balance with cell contents. Instead of single cells, the experiment uses kumara cylinders, each of which contains thousands of cells. Also use Biology 2.1, Unit 1 as a guide.

4

ISBN: 9780170372855

Enzymes

Thousands of different chemical reactions go on simultaneously inside any active cell. All this biochemical activity is together known as **metabolism**. 'Building up' reactions are **anabolism** and 'breaking down' reactions **catabolism** (from Greek, *ana* = up, *kata* = down).

Most metabolic reactions could not occur without enzymes. Enzymes are catalysts; they greatly speed up chemical reactions without being used up. Some enzyme molecules can be reused more than 10,000 times per second.

Almost all enzymes are proteins, made of long polypeptides (chains of amino acids) folded into specific three-dimensional shapes. The molecules on which an enzyme works are the **substrates**. Each enzyme is very specific: it can only act on a 'correct' substrate. **Co-factors** are substances that some enzymes need to work. Some are inorganic ions (like Cu^{2+}); others are organic molecules known as **co-enzymes**. Most co-enzymes come from vitamins.

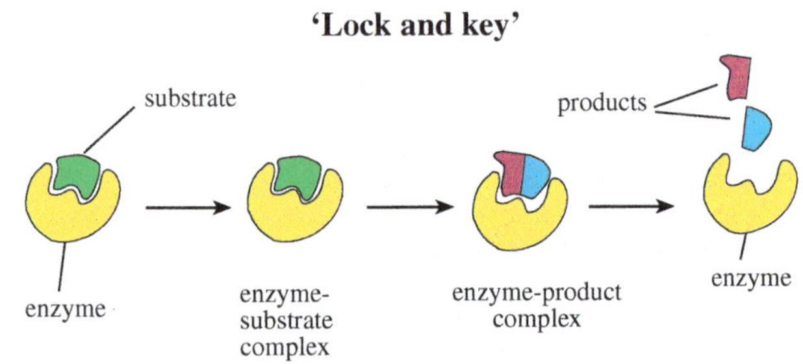

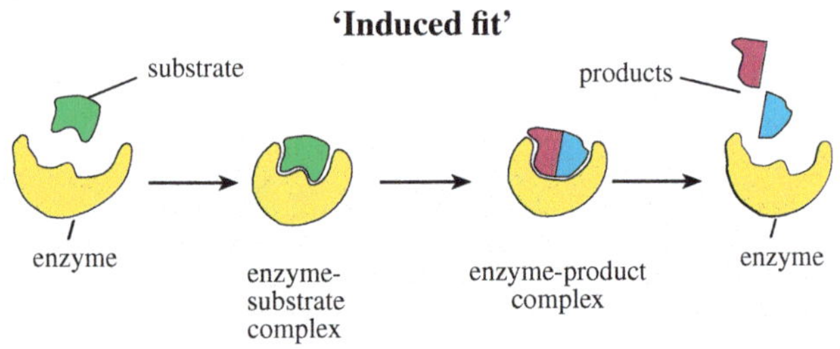

Fig. 2.4.14 Two slightly different hypotheses on how enzymes act on their substrate(s): lock and key assumes precise fit is needed; the more recent 'induced fit' idea is that the enzyme molecule is flexible enough to change its shape to fit the substrate molecule, altering the substrate in the process.

4

Enzymes and temperature

When temperatures rise, molecules collide more violently, which causes chemical reactions to go faster. With enzyme reactions, this rate increase stops beyond a certain temperature. Reason: higher temperatures cause proteins to unfold and change their shape. This destructive damage to enzymes is called **denaturing**, and explains why body temperatures above 42 °C are fatal.

An **optimal** temperature is one at which an enzyme works fastest. For many human enzymes this is about 37 °C. The optimum for prokaryotes living in hot springs can be 90 °C; in some cold-water fish the optimum may be as low as 10 °C.

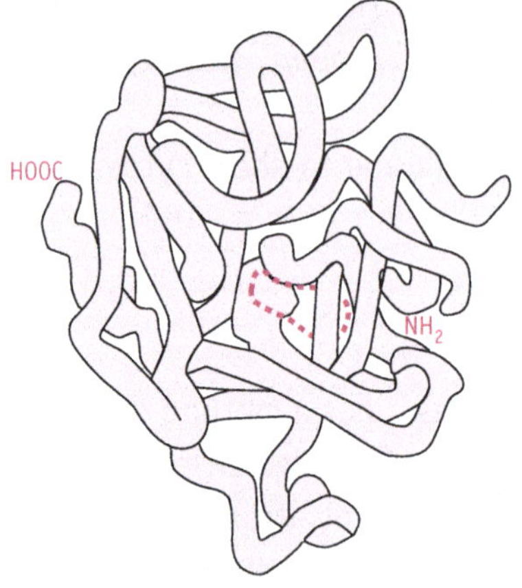

Fig. 2.4.15 Chymotrypsin, a protein-digesting enzyme produced in the pancreas. The position of the active site is shown in the red dotted line. The two ends of the polypeptide chain are represented by NH_2 and HOOC.

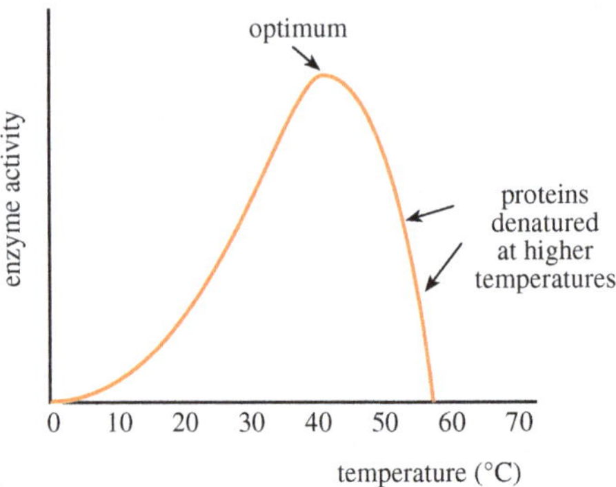

Fig. 2.4.16 The effect of temperature on the activity of pepsin, a digestive enzyme active in the stomach.

ISBN: 9780170372855

Enzymes and pH

The pH in cells is about 7.2 and many enzymes work best at close to this pH. However not all enzymes have an optimal pH of 7.2, as the graph shows.

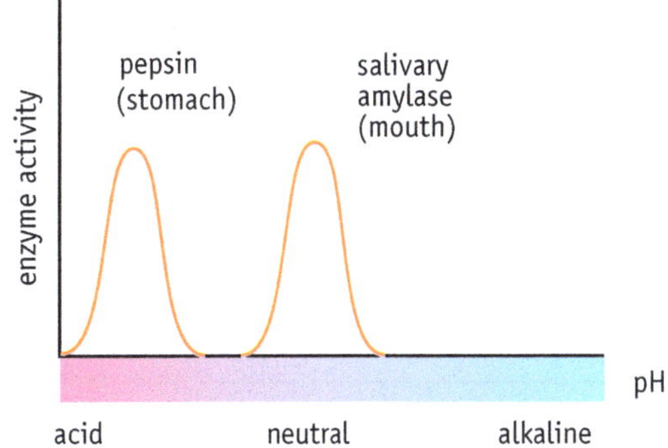

Fig. 2.4.17 The effect of pH on the activity of two digestive enzymes in humans. (These enzymes are active outside cells.)

E

Enzymes and substrate concentration

An enzyme can only work when substrate molecules collide with the active site. If there are few substrate molecules nearby there can be a delay between the departure of the product and the arrival of the next substrate molecule, and during this time the enzyme is kept waiting. At high substrate concentrations there may be no waiting time, so the enzyme cannot work any faster — it has reached its limit. Refer to Unit 5 information on **limiting factors** (page 81).

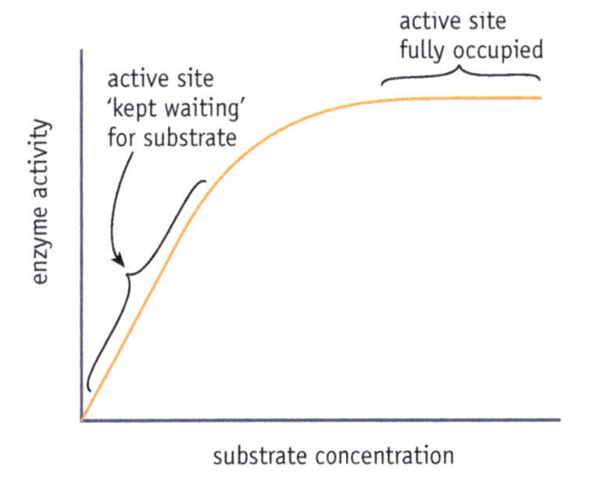

Fig. 2.4.18 The effect of substrate concentration on enzyme activity. Any further increase in substrate concentration will make no difference to reaction rate.

Activity B

1 Describe what happens when a protein molecule is denatured.

2 Suggest why protein molecules are easily denatured by temperatures as low as 45 °C, but sugar molecules are not affected.

ISBN: 9780170372855

3 Write the word(s) that match each definition or description in the table below. Select from this list: *pepsin, catalyst, metabolism, substrate, enzyme, optimal, induced fit, denatured, active site, product, catabolic, anabolic.*

a	A substance that speeds reactions without being used up	
b	A biological catalyst	
c	The active region on an enzyme molecule	
d	The substance(s) that an enzyme acts on	
e	The result of an enzymatic reaction	
f	A general word for chemical reactions in cells	
g	Processes that build big molecules from smaller ones	
h	Conditions under which a reaction works fastest	
i	A protein that has permanently changed its shape	
j	An example of a digestive enzyme	
k	Hypothesis about enzyme action	
l	Processes that break down big molecules	

4

Hands-on experiment: enzyme activity rate

Biology 2.1, Unit 2, Activity D describes a way of measuring the rate at which an enzyme breaks down hydrogen peroxide, enabling you to investigate factors that could influence enzyme activity rate. Investigate one of:

1 Temperature, including optimal temperature. Range suggested: approximately 20, 30, 40, 50, 60 °C.

2 pH, including optimal pH. Range suggested: pH approximately 4, 5, 6, 7, 8, 9. Your teacher will provide suggestions on how to achieve these settings for the yeast cells.

Start with a hypothesis that can be tested, and also use the experiment planning guidelines provided in Biology 2.1, Unit 1.

ISBN: 9780170372855

Unit 4 | DNA and cell division

Cells divide rapidly before birth and during early childhood. Later in life, cell replacement becomes the main reason for cell division. We say 'cell division', but the end result is actually multiplication.

Cell division involves:

- **Replication of DNA**, which starts long before the cell divides
- Division of the nucleus by **mitosis**, followed by division of the rest of the cell.

DNA's role and structure

There are many thousands of different kinds of protein, ranging from enzymes to polypeptide insulin. Each protein consists of one or more polypeptides, and each of these is a chain of amino acids linked together in a specific sequence.

The 'how to' information for assembling proteins is stored in genes. A gene is a length of DNA. Each chromosome contains hundreds or even thousands of genes. Humans have 23 pairs of chromosomes in almost every body cell.

DNA's unique roles are to:

- **carry coded information.** This information is carried in chemical form to regulate the making of proteins. (But DNA itself is not a protein!)
- carry genetic information **from one generation to the next.**

DNA is a polymer (a molecule with repeated components), made of sub-units called **nucleotides**. A DNA nucleotide has three parts:

- one phosphate group (P)
- one deoxyribose sugar (S), with five carbon atoms
- one base (B). There are four kinds of base: **adenine**, **guanine**, **cytosine**, **thymine**. (RNA, a similar molecule, has the base **uracil** in place of thymine.) Adenine and guanine are two-ringed **purines**; cytosine and thymine are one-ringed **pyrimidines**.

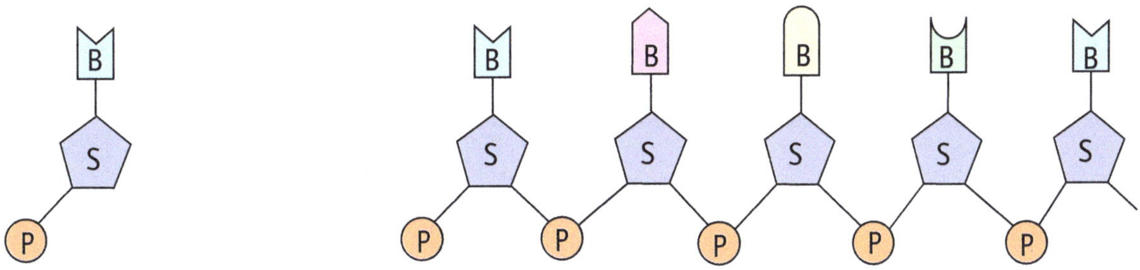

A single nucleotide

A short section of polynucleotide chain

Fig. 2.4.19 Left: a single nucleotide. Right: five nucleotides joined to make a short section of DNA.

Seen on a bigger scale, each DNA molecule consists of two polynucleotide chains wrapped around each other like a twisted ladder. The 'rungs' are pairs of bases, each pair linked by weak hydrogen bonds.

4

ISBN: 9780170372855

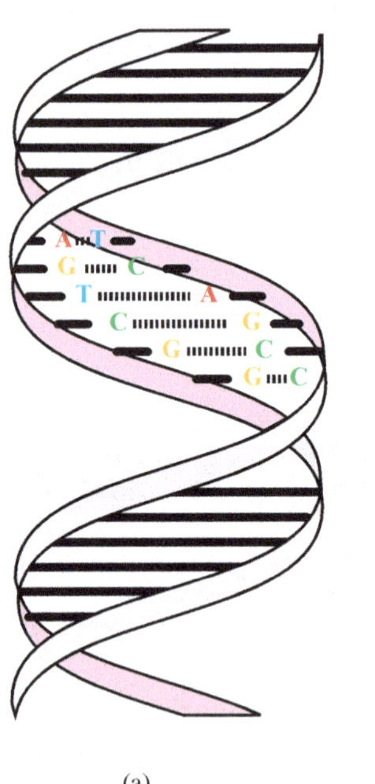

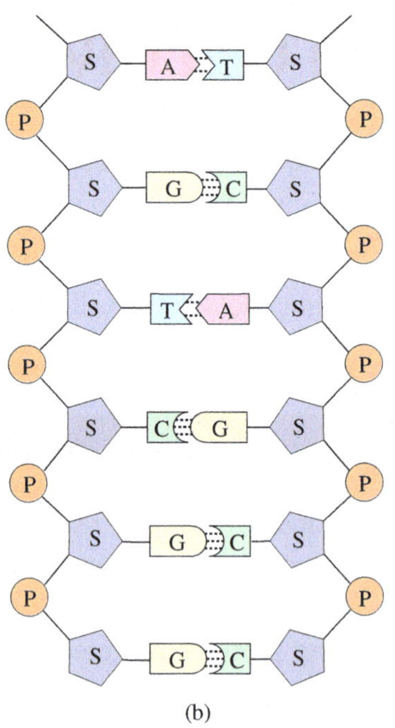

Fig. 2.4.20
(a) A simple representation of the DNA double helix.
(b) A helix straightened out to show base pairing between the two chains.

Pairing between bases follows strict rules:

- a **purine** always pairs with a **pyrimidine**
- **A** (**adenine**) always pairs with **T** (**thymine**)
- **G** (**guanine**) always pairs with **C** (**cytosine**)

These 'rules' result in **complementary base pairing**, which means that if the base sequence in one strand is known, the base sequence in the other can also be known.

Example: if the sequence on one strand is T T A C G G C A A T G G C,
then the sequence on the other strand must be A A T G C C G T T A C C G.

DNA replication

Before a cell starts to divide, all the DNA in its chromosomes is copied. Copying is so accurate that the two resulting cells and all their descendants have exactly the same sequence of bases. In the case of humans, this means copying a sequence of about three billion base pairs.

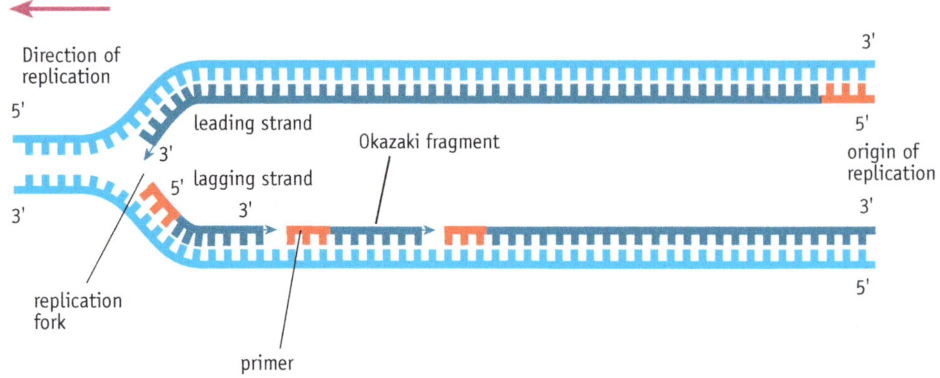

Fig. 2.4.21 Replication of DNA adds nucleotides at the growing ends of new strands, but only at 3′ ends. As a result one strand grows continuously and its opposite strand in sections (Okazaki fragments). Not shown here: A-T and C-G base pairing.

ISBN: 9780170372855

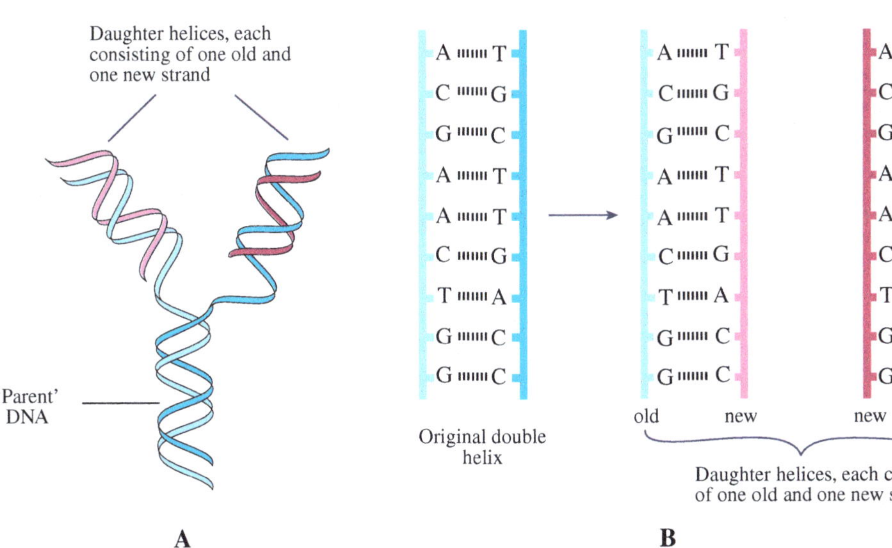

Fig. 2.4.22
A DNA replicating, showing no details of base pairing. New strands are shown in red and pink.
B Double strand straightened out, before and after replication.

After replication two identical DNA molecules are produced, each a copy of the original DNA. Each 'daughter' DNA molecule consists of one old and one new strand. For this reason, DNA replication is said to be **semi-conservative** (conservative meaning 'keeping the same').

The processes of DNA replication and mitosis are the same in male and female, but it's customary to label the original cell (or DNA strand) the 'parent', and the two resulting cells (or DNA molecules) the 'daughters'. Daughter cells are genetically identical to each other, and also identical to the parent cell.

4

Activity A

1 Draw a single labelled nucleotide containing thymine.

2 Describe the base pairing rule.

____

____

3 Explain what is meant by a 'polymer'.

____

____

4 In one word, give the name for the 'monomer' units of DNA. ______________________

ISBN: 9780170372855

5 Summarise the function (purpose) of DNA replication.

6 Explain what is meant by DNA replication being 'semi-conservative'.

7 This represents one side of the DNA strand: T T G A G C A T A C C G.

Write the complementary bases here: ______ ______ ______ ______

8 The following events A–F are all part of DNA replication, but listed here in the wrong order. Decide on the correct order and write the matching letters on this line: ______

A the original DNA unwinds and unzips

B each double strand now forms a helix

C hydrogen bonds between paired bases are broken by an enzyme

E adjacent nucleotides link up

D new nucleotides are brought into position

F new and old nucleotides pair up

9 Arrange the following six structures in order of size, starting with the smallest: DNA double helix, nucleotide, base, nucleus, cell, chromosome.

10 Evaluate the significance of DNA base pairing, in relation to DNA's overall function.

11 Describe what is meant by 'Okazaki fragments'. Explain why two complementary strands of DNA grow in different ways during replication.

4

ISBN: 9780170372855

Mitosis

Chemically, a chromosome is a DNA strand supported by a protein framework. Humans have 23 pairs of ribbon-like chromosomes in each body cell, but only four pairs are shown in Fig. 2.4.23.

Mitosis is the process of changes that happen in a nucleus during cell division. **Meiosis** is different. This is a more complex kind of nuclear division that forms part of sexual reproduction, and is described in Biology 2.5. It's important not to confuse mitosis and meiosis.

Cell division: where and when

Mitosis happens most often during embryo development. Starting from one single cell nine months earlier, a human baby is born with about 10^{13} cells. Some tissues are continuously being worn away, such as skin and the lining of intestines, so continue to make new replacement cells throughout life. In tissues like muscles and brain, comparatively few new cells are made over the rest of life.

Mitosis stages

Mitosis is a continuous process, but is for convenience divided into named stages. You do not need to learn the stage names; what matters most is chromosome numbers and behaviour. Numbers 1 to 8 refer to individual cell drawings in Fig. 2.4.23.

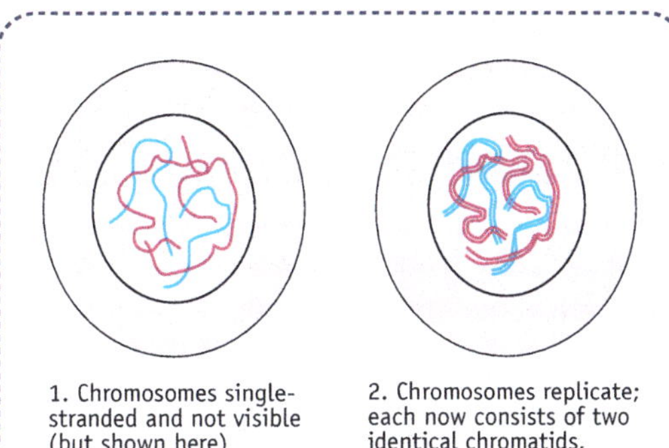

1. Chromosomes single-stranded and not visible (but shown here)

2. Chromosomes replicate; each now consists of two identical chromatids, shown but still not visible

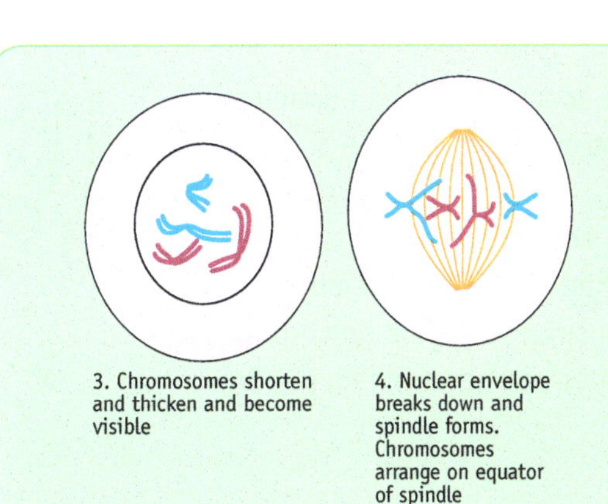

3. Chromosomes shorten and thicken and become visible

4. Nuclear envelope breaks down and spindle forms. Chromosomes arrange on equator of spindle

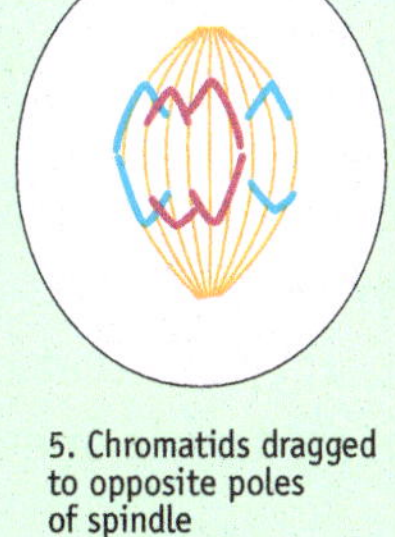

5. Chromatids dragged to opposite poles of spindle

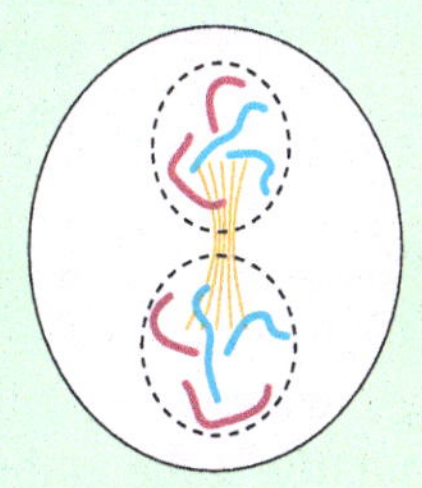

6. Each pole of the spindle receives an identical set of daughter chromosomes

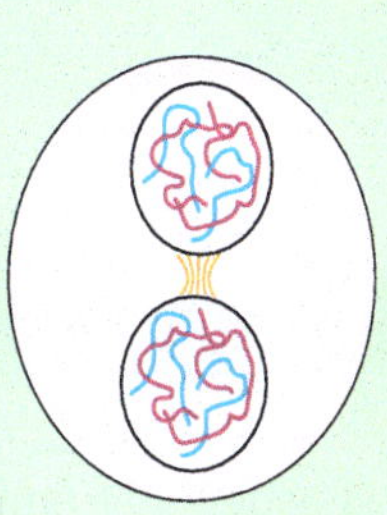

7. Chromosomes lengthen and become indistinct; nuclear envelope forms round each chromosome cluster

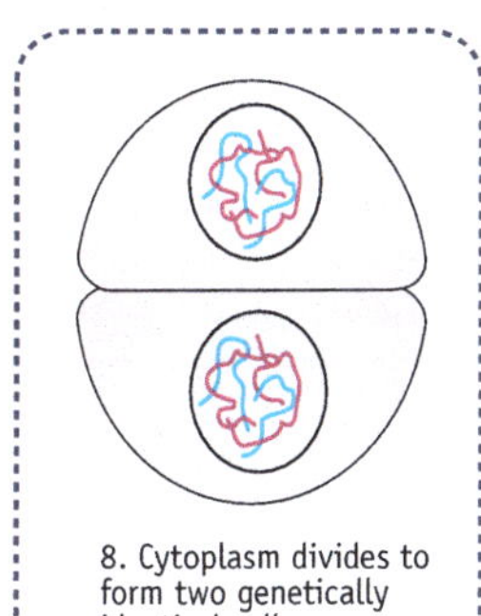

8. Cytoplasm divides to form two genetically identical cells

Fig. 2.4.23 Cell drawings 3–7 in the green frame show different stages of mitosis. Stages 1, 2 and 8 shown in the dotted frames occur before and after mitosis.

1 and **2** (Interphase) This occurs before mitosis starts. DNA is replicated, but this is not yet visible. Each chromosome now consists of two identical chromatids.

3 (Prophase) The chromosomes become shorter and more clearly visible. The nuclear envelope breaks down and a spindle develops — a system of protein threads. In animals and in some plants, this is organised by the centrioles.

4 (Metaphase) The centromere of each chromosome attaches to the 'equator' of the spindle.

5 (Anaphase) The two chromatids of each chromosome are pulled apart to opposite ends of the spindle, and are now called daughter chromosomes. Each end thus receives one of the two daughter chromosomes formed from each original chromosome.

6 and **7** (Telophase) A nuclear envelope develops around each cluster of chromosomes, forming two genetically identical nuclei.

8 (Cytokinesis) Cell divides more or less equally, forming two genetically identical cells.

E

Why do cells divide?

The need for oxygen and raw materials depends on the volume of the cell, but supply depends on the cell surface area. When a cell grows and doubles in width, its volume becomes eight times as much as before (in proportion to d^3), but its surface area only increases four times (in proportion to d^2). Smaller objects have greater surface area-to-volume ratios compared with large objects of the same shape. Cells divide because it keeps the surface area big enough for oxygen to diffuse into tissues fast enough.

The cell cycle

Cell division is very rapid in a growing embryo. Each daughter cell follows one of two paths:

- It may go on to divide again — this is what stem cells do.
- Or it may become specialised for a particular function, and not divide again.

In certain parts of an adult body, cells continue to divide. Example: skin. Whether in embryo or adult, any cells that continue to divide go through a repeated cycle of events, as shown in Fig. 2.4.24.

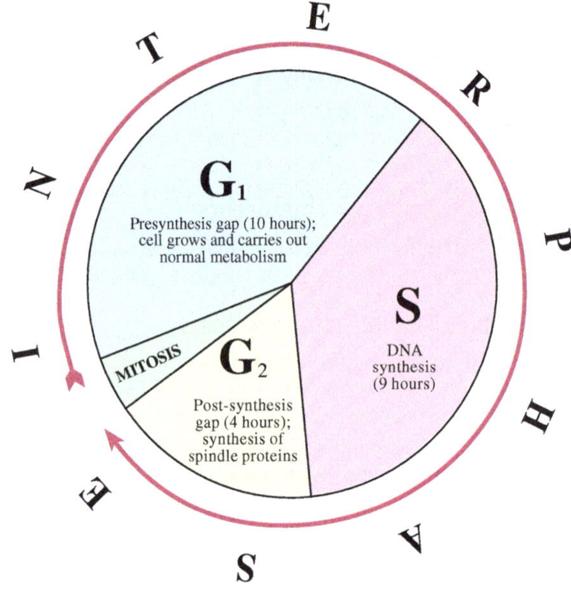

Fig. 2.4.24 The cell cycle. Times are for human cells grown in artificial conditions. Mitosis is quite brief. Most of the cycle is interphase, with no visible changes to the chromosomes.

Activity B

1 Explain why mitosis is especially rapid in early-stage embryos.

ISBN: 9780170372855

2 Name one kind of cell in adults where mitosis frequently occurs, and explain why these cells keep dividing.

____

____

3 Explain the difference between 'chromosome' and 'chromatid'.

____

____

4 Explain the difference between 'centriole' and 'centromere'.

____

____

5 Explain what is meant by 'daughter cells'.

____

____

6 The following questions refer to the numbers in a normal human body cell.

a How many chromosomes are in a normal human body cell? _______________

b How many chromatids are in a nucleus at prophase stage of mitosis? _______________

c How many chromosomes are in each nucleus just before cytokinesis? _______________

d How many chromosomes are in each cell after cytokinesis? _______________

7 The following 10 events A–J are all part of mitosis and cell division in animal cells, but are listed here in the wrong order. Decide on the correct order and write the matching letters on this line, from first event to last.

____

A spindle fibres pull chromatids apart
B chromatids shorten and thicken
C a spindle forms, developed from the centrioles
D nuclear envelope(s) form around the two groups of chromosomes
E cytokinesis; the cell divides in two parts
F DNA replicates
G centromeres attached to the spindle fibres
H nuclear envelope disappears
I each chromosome is replicated, forming two chromatids
J chromatids form two groups at the ends of the cell

ISBN: 9780170372855

Unit 5 | Specialised cells

Cells differ greatly in shape and size and internal details, the differences being due to the functions of specialised tissues.

Some tissues form layers of cells (e.g. epithelium); mostly they are not in layers (e.g. muscle tissue). The word 'tissue' is used for both plants and animals.

> **Tissue**: a number of similar cells acting together for a special purpose.

There are many different kinds of tissue, with functions as varied as:

- secretion, which means producing a substance for a particular purpose
- protection
- storage of energy-rich substances
- contraction (muscles).

To explain how cells or tissues are specialised for their particular function, we can look at:

1 cell shape; e.g. increased surface area for absorption
2 cell size; very small cells have a proportionately larger surface area
3 cell wall thickness (only in plants)
4 abundance of particular organelles; especially mitochondria
5 organelle placement; e.g. chloroplasts are placed to capture CO_2.

4

Some animal cell types

Example

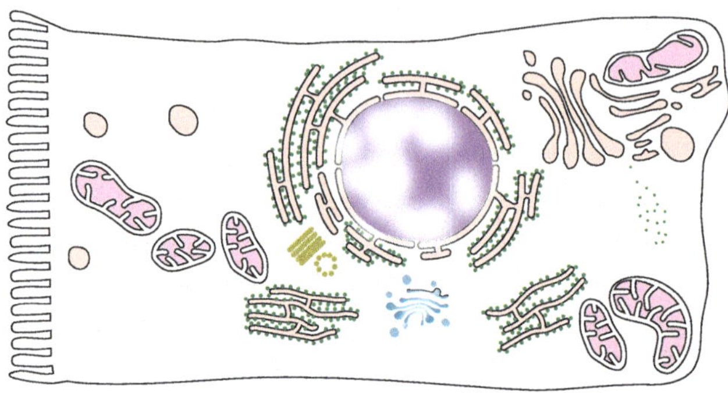

Fig. 2.4.25 Cell from tissue lining the small intestine, specialised for food absorption. The plasma membrane is extended into many finger-like projections and the cell has many mitochondria.

Feature	**How?** (how it works) and …	**Why?** (the function)
Many small projections	This increases the surface area …	… which increases the overall rate of food absorption.
Many mitochondria	Absorption of food into blood is 'uphill', so active transport is involved …	… and this needs ATP energy supplied by mitochondria.

ISBN: 9780170372855

Activity A

For each of the four animal cell types drawn below, explain how the identified feature relates to the overall function of that cell or tissue.

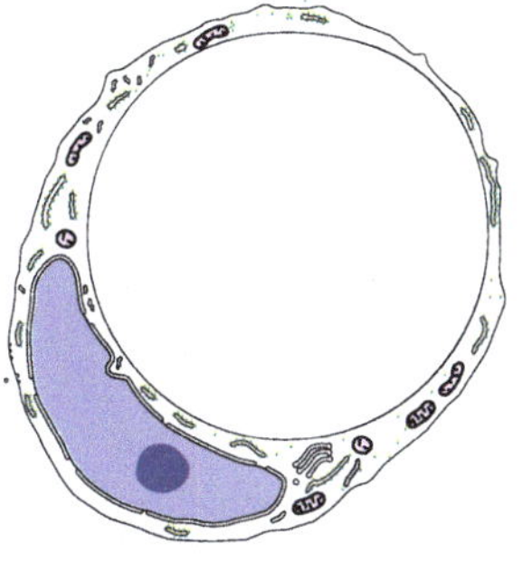

1 Fat cell

Feature: large lipid droplet

Explain ____________________

____

____

Feature: few mitochondria

Explain ____________________

____

____

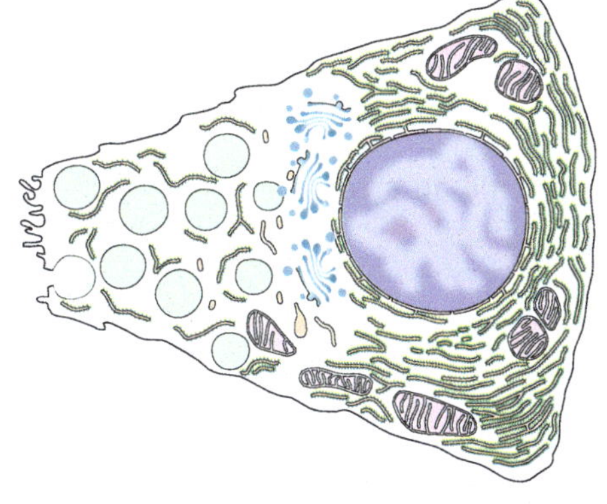

2 Secretory cell in salivary gland

Feature: many RER and Golgi complexes

Explain ____________________

____

____

Feature: many mitochondria

Explain ____________________

____

____

Feature: several vacuoles

Explain ____________________

____

____

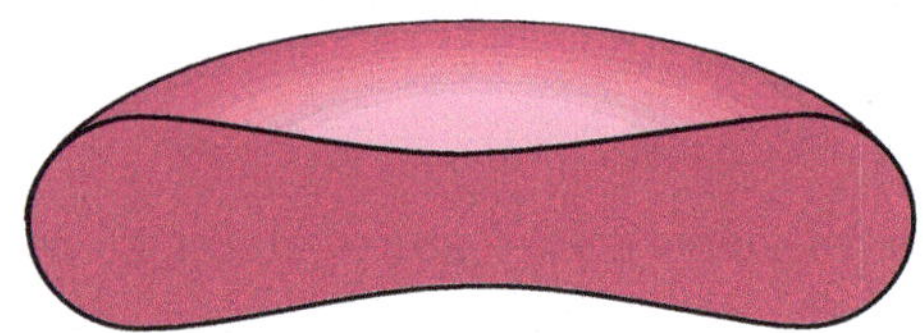

3 Red blood cell

Feature: very small size

Explain ______________________________

____

____

Feature: filled with haemoglobin

Explain ______________________________

____

____

Feature: no organelles

Explain ______________________________

____

____

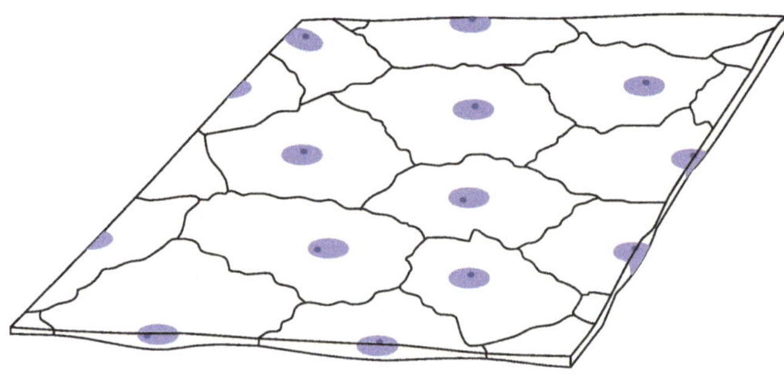

4 Epithelial cells lining an artery

Feature: thin flats cells

Explain ______________________________

____

____

Feature: bound tightly together

Explain ______________________________

____

____

ISBN: 9780170372855

Activity B

For each of the three plant cell types drawn below, explain how the identified feature works (its mechanism), and also how it relates to the overall function of that cell or tissue.

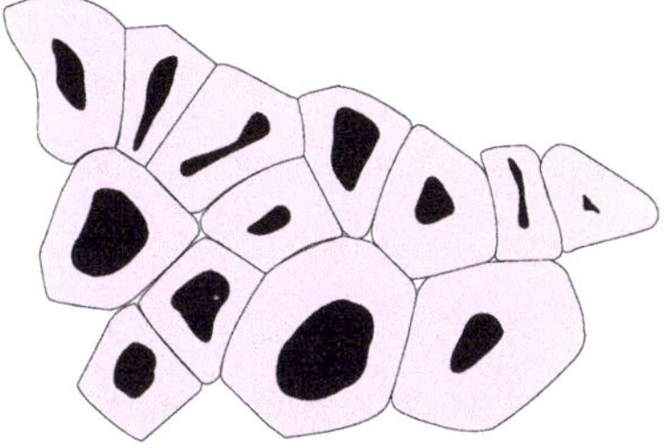

1 Cells in the woody part of a tree

Feature: thick cell walls

Explain ______________________________

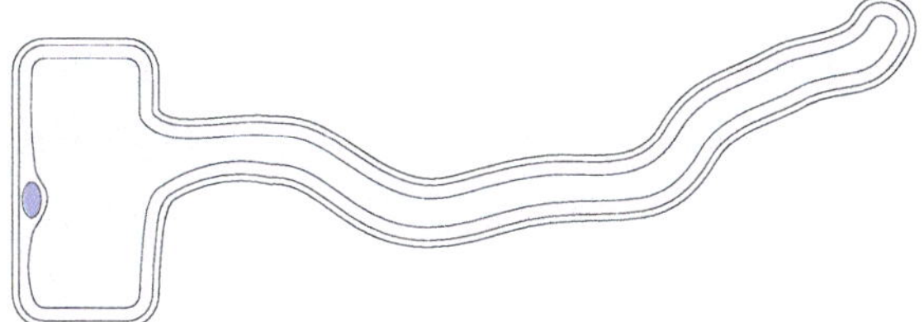

2 Hair-shaped cell on surface of young plant roots

Feature: long thin shape

Explain ______________________________

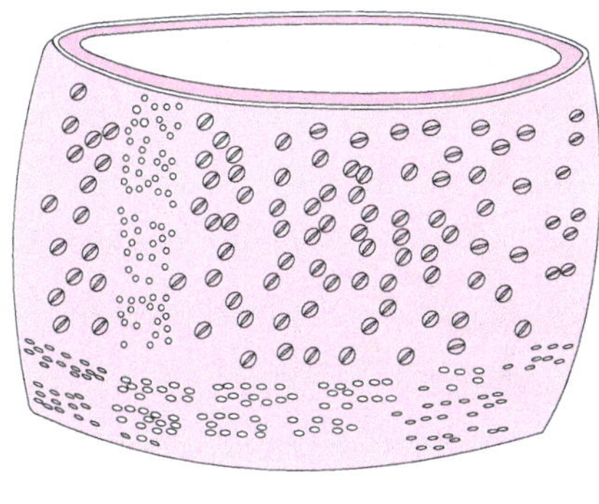

3 Xylem cell for carrying water

Feature: shape suited for transport

Explain ______________________________

Feature: no end walls

Explain ______________________________

ISBN: 9780170372855

Leaf structure

Most kinds of leaf are specialised for photosynthesis. (In some plants, leaves have been reduced to sharp spikes for protection from animals, and are not involved in photosynthesis.) To help the process go as fast as possible, leaves have features that:

- maximise the amount of light absorbed
- maximise the rate of CO_2 uptake (important, since air is only 0.04% CO_2)
- get water (via veins)
- export the sugar they produce.

Reminder: photosynthesis is a complex series of reactions that can be summed up in a simple equation:

$$CO_2 \text{ (from the air)} + H_2O \text{ (from the soil)} + \text{light} \rightarrow \text{sugars} + O_2$$

Chlorophyll (in the chloroplasts) is an essential part of the process.

4

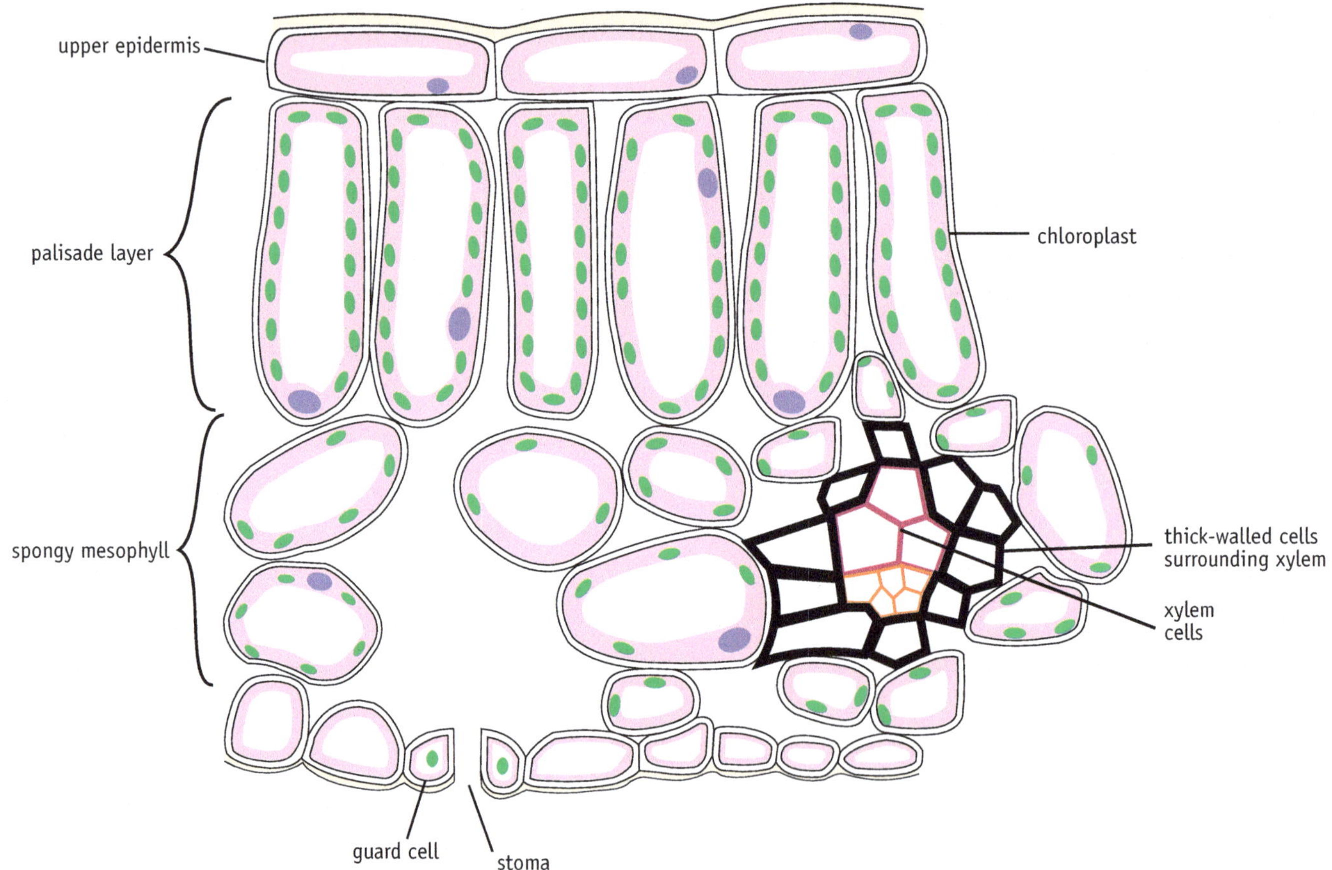

Fig. 2.4.26 Cross-section of a 'typical' leaf to show some features that increase the rate of photosynthesis. (The drawing is simplified. In reality most cells are much smaller, in relation to leaf thickness.)

ISBN: 9780170372855

Limiting factors

There is a limit to how fast photosynthesis can work. One factor always limits the process; either light availability, or CO_2 absorption, or enzyme effectiveness, or some other factor. The situation is summed up by the graph curves in Fig. 2.4.27.

The Principle of Limiting Factors: In any process that involves a number of raw materials, the overall rate will be limited by whichever raw material is in shortest supply.

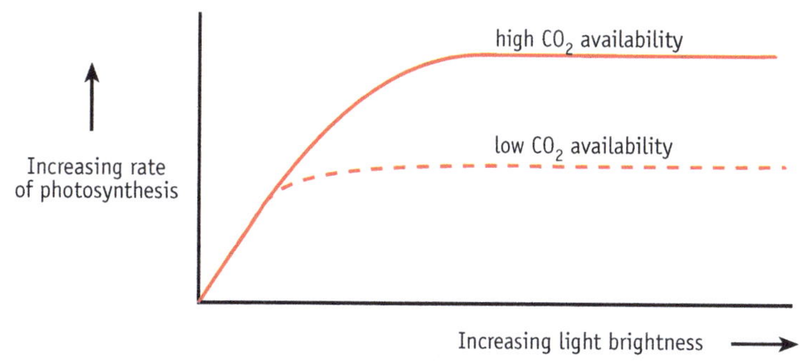

Fig. 2.4.27 The Principle of Limiting Factors, seen here affecting photosynthesis.

Activity C

For each of features **1** to **7** listed below, explain *how it works* (its mechanism) and also *why it exists* (its function). Refer to Fig. 2.4.26.

1 Upper and lower epidermis cells have no chloroplasts. (Except for guard cells.)

2 Palisade cells are oriented at 90° to the surface.

3 Many chloroplasts in the palisade cells, compared to elsewhere.

4 There are thousands of stomata per square cm.

ISBN: 9780170372855

5 Leaf interior (mesophyll) is spongy, with many air spaces.

6 The veins have xylem cells, which carry water.

7 Thick-walled cells surrounding the veins.

8 Use the Limiting Factor Principle to explain why plants given 10 times the light intensity will not grow 10 times as fast.

9 Refer to Fig. 2.4.27.

a Explain why photosynthesis is faster when more CO_2 is available.

b Suggest why the graph 'levels off' even when more CO_2 is available.

4

ISBN: 9780170372855

Biology 2.4 Cell biology

AS 91156 Demonstrate understanding of life processes at the cellular level
Externally assessed, 4 credits

Achievement	Achievement with Merit	Achievement with Excellence
Demonstrate understanding of life processes at the cellular level.	Demonstrate in-depth understanding of life processes at the cellular level.	Demonstrate comprehensive understanding of life processes at the cellular level.

Achievement
'Demonstrate understanding ...' involves defining, using annotated diagrams or models to describe, and describing characteristics of, or providing an account of, life processes at the cellular level.

Achievement with Merit
'Demonstrate in-depth understanding ...' involves using biological ideas to give reasons how or why life processes occur at the cellular level.

Achievement with Excellence
'Demonstrate comprehensive understanding ...' involves linking biological ideas about life processes at the cellular level. Discussion of ideas may involve justifying, relating, evaluating, comparing and contrasting, analysing.

Notes

1 'Life processes at the cellular level' include:
- photosynthesis
- respiration
- cell division (DNA replication and mitosis as part of the cell cycle).

2 'Biological ideas' that relate to each of the life processes at the cellular level are selected from:
- movement of materials (including diffusion, osmosis, active transport)
- enzyme activity (specific names of enzymes are not required)
- factors affecting the process
- details of the processes only as they relate to the overall functioning of the cell (specific names of stages are not required)
- reasons for similarities and differences between cells such as cell size and shape, and type and number of organelles present.

Cells include plant cells and animal cells.

3 Examination: Candidates may be required to interpret diagrams and new information, draw diagrams, and write responses of one or more paragraphs. Some questions may be resource-based. Candidates are expected to understand the structure of DNA, and the meaning of semi-conservative replication. Factors affecting the processes may include both direct and indirect availability of resources for the plant or animal. Factors affecting enzyme activity within cells may include temperature, pH, substrate concentration, enzymes, enzyme inhibitors. Similarities and differences between cells may relate to the overall functioning of the organism, and justifying the reasons for the similarities and differences. Movement of materials may also include facilitated diffusion.

ISBN: 9780170372855

Revision 1 | Cell biology 2.4

Write a list of key points in each box.

Organelles

Transport

Enzymes

DNA replication and mitosis

ISBN: 9780170372855

Revision 2 | Cell biology crossword

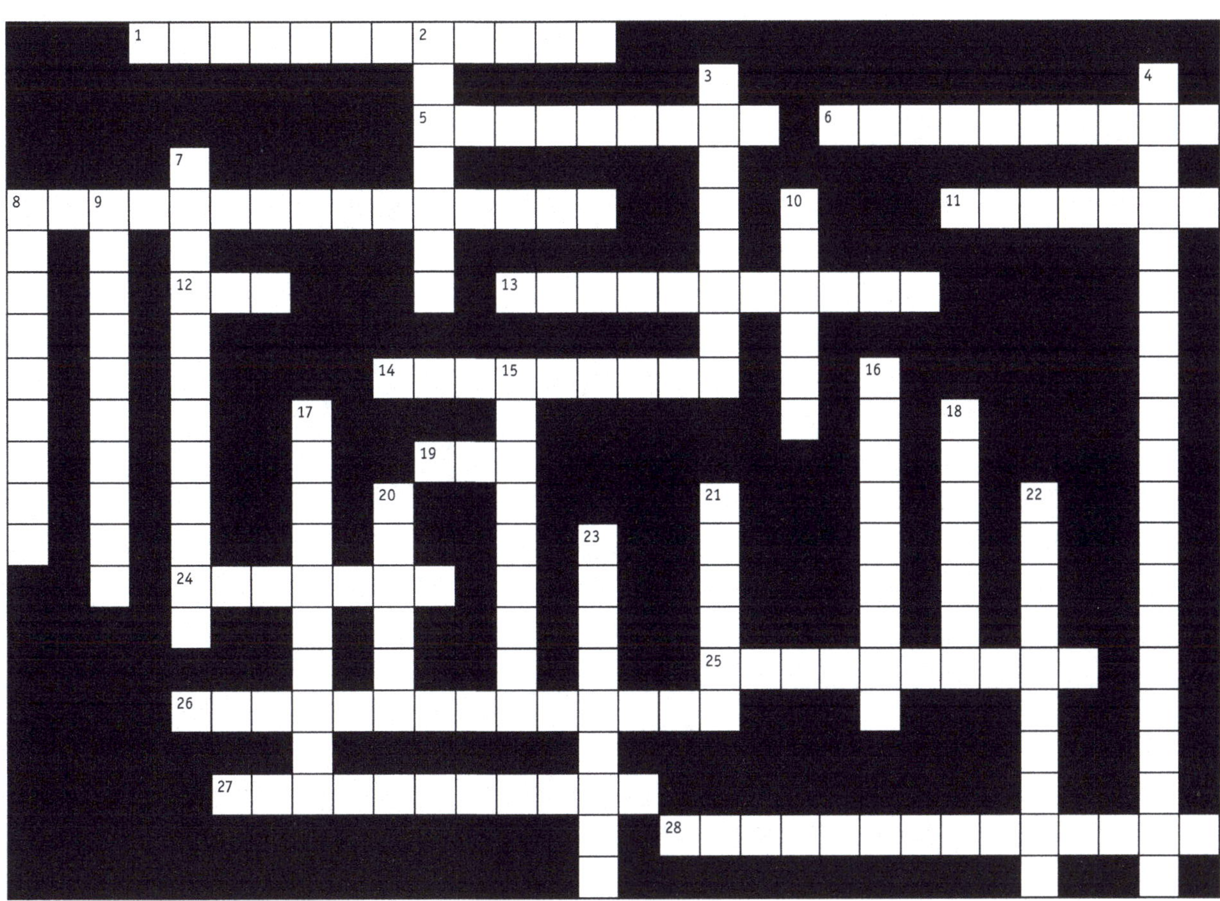

Across

1 Weak force between complementary bases in DNA (8, 4)
5 Product of chromosome replication (9)
6 Single-ringed base in DNA (10)
8 Separates nucleus from cytoplasm (7, 8)
11 Watery solution in which cytoplasmic organelles are suspended (7)
12 Immediate source of energy in cells (3)
13 The site of photosynthesis (11)
14 Membrane-bound region of a cell specialised for a particular function (9)
19 The material of which genes are made (3)
24 Movement of water across a semi-permeable membrane (7)
25 A millionth of a metre (10)
26 Outer boundary of the cytoplasm (6, 8)
27 Process in which mitochondria make most of a cell's ATP (11)
28 Organelle in which newly synthesised proteins are further processed (5, 9)

Down

2 Organelle in which chromosomes are located (7)
3 Tiny granule on which polypeptides are produced (8)
4 Complex network of membrane-bound spaces in cytoplasm (11, 9)
7 Inactivation of enzyme by permanent change of shape of its active site (12)
8 A thousandth of a millimetre (9)
9 Threadlike structure containing many genes (10)
10 Organelle that produces movement (6)
15 Making large molecules from smaller ones (9)
16 Net movement of a substance from higher to lower concentration by random particle movement (9)
17 Breakdown of glucose to pyruvate plus a little ATP (10)
18 Large, permanent fluid-filled cavity in plant cell (7)
20 Fold in inner mitochondrial membrane (6)
21 Organic catalyst (6)
22 Part of a chromosome that attaches to the spindle in cell division (10)
23 Organelle that organises the spindle in animal cells (9)

ISBN: 9780170372855

Exam-type question 1

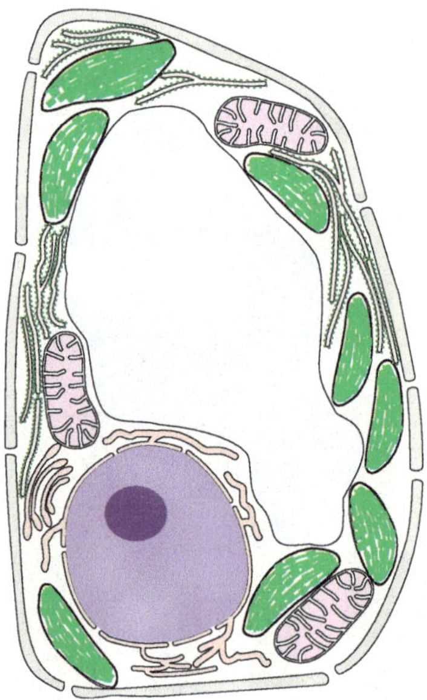

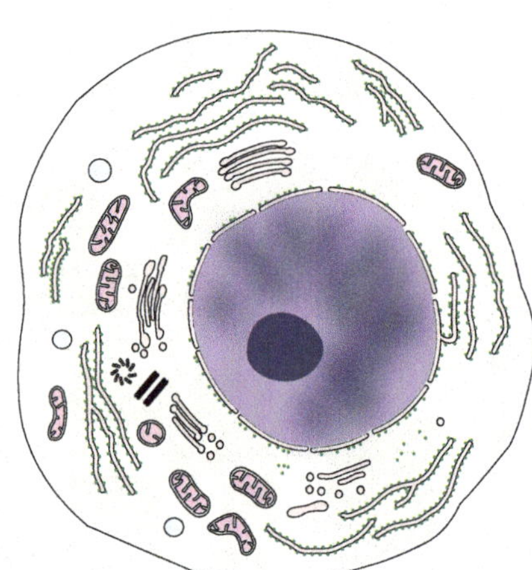

Two kinds of organelle shown in the cells above are the mitochondrion and chloroplast.

- Describe the cellular processes related to each organelle.
- Discuss the similarities and differences in structure of the two organelles.
- Explain how the structures of both organelles relate to their very different functions.
- Also explain where the organelles are positioned in the cell and why.

Use this scaffold framework as the basis for an answer to be done on your own paper.

- Specify what process occurs in mitochondria and in chloroplasts and include a word equation for each process.
- For these two organelles, deal with their structure/features in detail. State in what way they are **similar**, e.g. 'Both organelles are enclosed by plasma membranes; both have specialised functions; both have specialised enzymes; both are very small, approximately 1 or 2 micrometres long.'
- Explain briefly how some of these features link to processes going on in the organelles. Regarding membranes, state which materials move in and out of the organelle due to the cellular process carried out in that organelle.
- Now discuss **differences** in the structure/features of chloroplasts and mitochondria. Examples: chlorophyll, thylakoids, stroma, cristae.
- The final part of the question requires you to explain **where** mitochondria and chloroplasts are situated inside cells.
- Suggest what benefits there could be to being close to the edge, compared to the middle of the cell.
- Refer to the diagrams above to justify your reasons with specific examples.

ISBN: 9780170372855

Exam-type question 2

Plant leaves are specifically adapted in a number of ways to ensure photosynthesis occurs efficiently. Study this diagram of a generalised leaf.

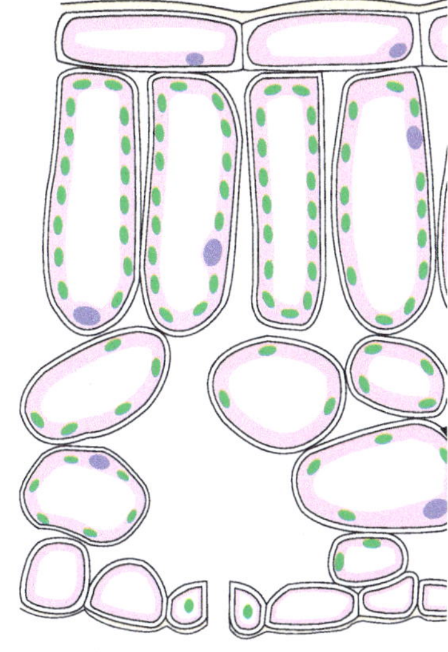

Choose four adaptations visible in this cross-section diagram. Explain and discuss how the structure, position or shape of each contributes to increasing the rate of photosynthesis. You may choose to mention:

- the position and number of chloroplasts
- the palisade and spongy layer
- the stoma and guard cells
- the thickness of leaf
- contents of epidermal cells.

Use this page to create a key points plan or a mind map or scaffold that could form the basis for a longer answer to be done on your own paper.

4

ISBN: 9780170372855

Exam-type question 3

Study the diagram and discuss the replication of DNA by explaining the following points:

- Explain how the process occurs (names of enzymes are not necessary).
- Explain how the process ensures that replication is accurate.
- Explain the term 'semi-conservative' as it applies to DNA replication.

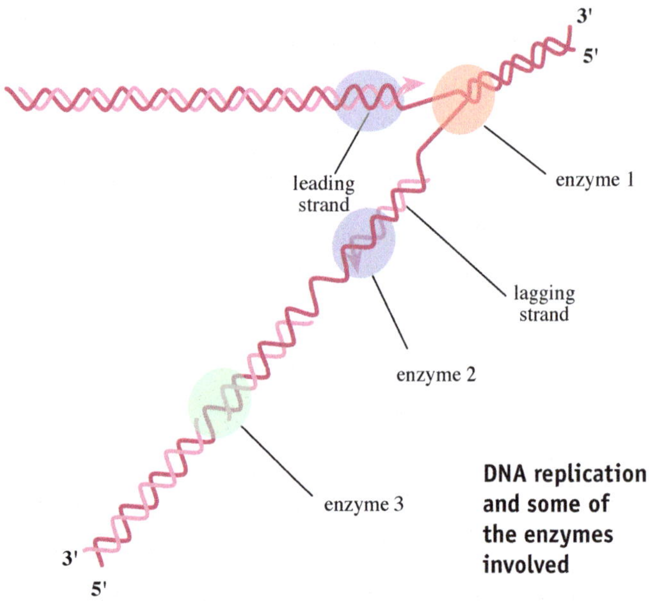

DNA replication and some of the enzymes involved

Use this scaffold framework as the basis for an answer to be done on your own paper.

- Write a step-by-step sequence of how DNA is copied, starting with how DNA unzips.
- Describe how the new bases match up with the open strands.
- Use technical words: template, nucleotides, complementary, exposed bases, enzymes.
- Explain why the two sides do not proceed at the same rate.
- Use technical terms/words such as: anti-parallel, 3′ end, Okazaki fragments, enzyme.
- Make sure you state how the process ends.
- Now explain how the copying is kept accurate/faithful.
- Use words such as: template, complementary base pairs, H bonds between bases, named enzymes that are involved.
- Lastly explain the term 'semi-conservative'.
- Make a drawing if it assists your explanations.

 ISBN: 9780170372855

Exam-type question 4

Humans have complex metabolism under control of the central nervous system and also hormones. Adrenaline is produced as a 'fight/flight' hormone in times of stress, to allow the body to deal with or avoid danger. It is manufactured in the adrenal glands and is released into the bloodstream; its effects include the rapid release of sugar for respiration. Three organelles work together to produce the product adrenaline: the nucleus, rough endoplasmic reticulum, and the Golgi apparatus.

Outline how these organelles function to produce adrenaline. Describe and explain:

- how nucleus DNA is involved in the process
- how the rough endoplasmic reticulum receives information to produce a product
- the role the Golgi apparatus plays in the process
- how the product will be transported out of the cell.

Use this page to create a key points plan or a mind map or scaffold that could form the basis for a longer answer to be done on your own paper.

ISBN: 9780170372855

Exam-type question 5

Below are diagrams of a root hair cell and a cell from a kidney tubule.

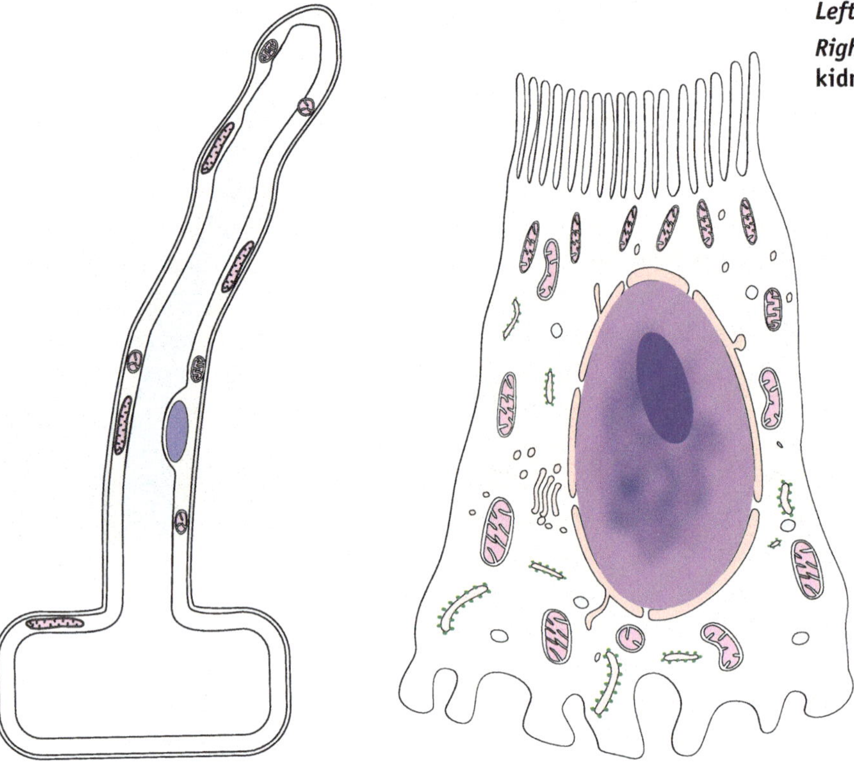

***Left*: A root hair cell.**
***Right*: A cell from a kidney tubule**

Study the diagrams carefully before answering the following questions.

Active transport and osmosis are two processes that occur in both types of cell.

Discuss these processes by addressing the following points.

- How the shape of both cells relates to transport of named substances specific to each cell type.
- Compare and contrast active transport and osmosis.

Use this scaffold framework as the basis for an answer to be done on your own paper.

- Describe first how the **shape** of each cell aids the rate of transport …
- … of named materials in and out of the cell.
- Define active transport and osmosis.
- Explain what these two processes have in common, e.g. 'Both processes involve …'
- Name specific molecules involved.
- Contrast the processes, i.e. point out differences between the two cells.
- You could use words such as: However, …, In contrast, …, Unlike …'
- Take one point of difference between the two processes and explain in greater detail.
- For example, start with the different materials transported.
- Compare molecule sizes, speed of transport, direction of movement, whether energy is needed or not, and why. Refer to passive/active transport.
- Also consider where exactly in the cell the transport occurs, whether any proteins are involved, and how the concentration of materials affects the rate of movement.

 ISBN: 9780170372855

Exam-type question 6

Plants are critical to almost all food chains. As the human population expands, we depend upon ongoing horticultural research to ensure that crop plants grow to their potential.

Photosynthesis requires enzymes. These have specific conditions that must be met in order for them to work to their maximum potential. Discuss how changing the following conditions can either improve or inhibit the rate of photosynthesis. Refer to enzyme activity in your answer, giving named examples of enzymes where possible.

- pH
- Substrate concentration
- Chemical herbicides

Use this page to create a key points plan or a mind map or scaffold that could form the basis for a longer answer to be done on your own paper.

ISBN: 9780170372855

Exam-type question 7

Below is a diagram of an organelle involved in photosynthesis.

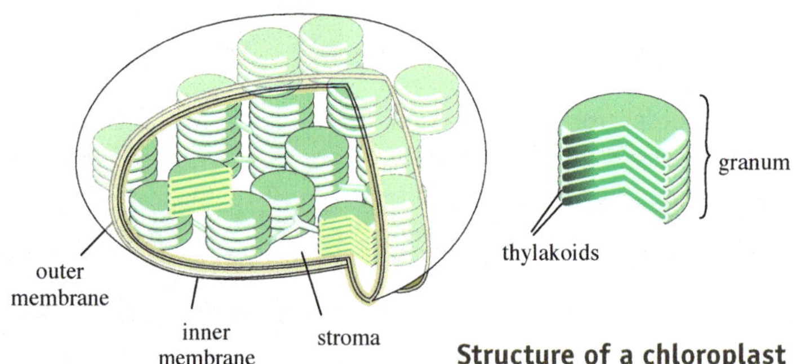

Structure of a chloroplast

Discuss how photosynthesis occurs in this organelle, by including:

- a description of the process.
- the specific structures involved.
- how the location of the organelle affects photosynthesis.
- how any two other factors affect the rate of photosynthesis.

4

Use this scaffold framework as the basis for an answer to be done on your own paper.

- Start with a description, e.g. 'Photosynthesis is the process in which …'
- Also write a word equation for photosynthesis.
- Use the diagram above to identify and describe the structures involved …
- … and state the role(s) they play in photosynthesis.
- Describe where chloroplasts are usually placed in plant cells …
- … and explain why they are placed there, relating this to information you have already given on the process of photosynthesis.
- Identify one factor that can speed up or slow down photosynthesis …
- … and explain how this speeding/slowing occurs, based on your knowledge of the steps involved in photosynthesis.
- Relate this speeding/slowing to specific named inputs and/or products.
- Identify a second factor that affects the rate of photosynthesis.
- Explain the speeding/slowing effect of this second factor in the same way as you dealt with the first factor.

 ISBN: 9780170372855

Exam-type question 8

The diagram below illustrates six stages in the mitotic cell cycle.

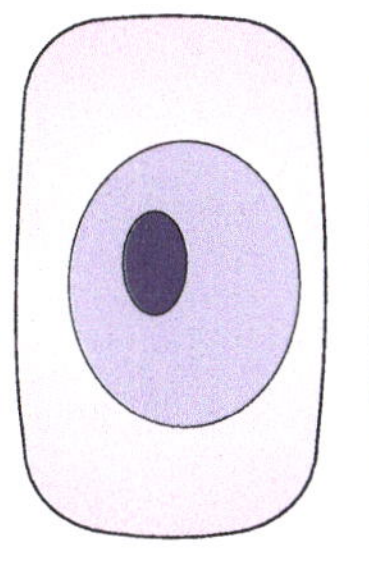
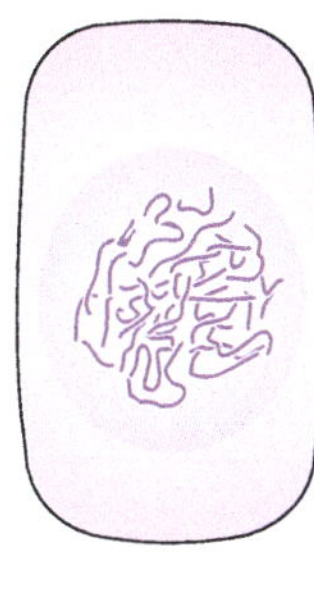

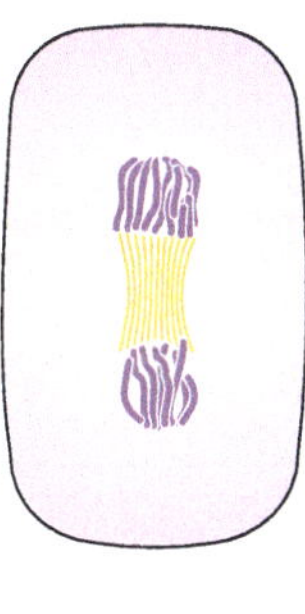
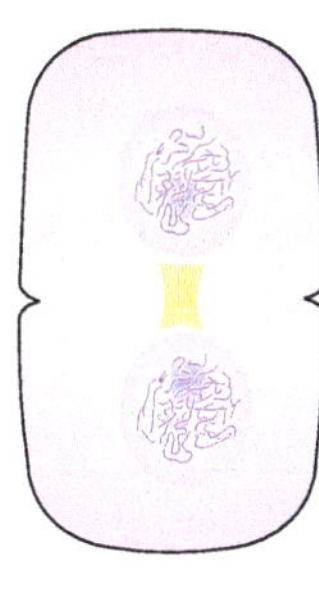

A B C D E F

An animal cell at various stages of the mitotic cell cycle

Mitotic cell division is vital to growth in both plants and animals, although the rate of cell division is not the same throughout the life of the organism.

Discuss this statement and include:

- a description of the process of mitosis (names of stages not required).
- a discussion of how the duration of the cell cycle might be expected to differ at different stages of life and in different parts of the body of a mammal and a flowering plant.

Use this page to create a key points plan or a mind map or scaffold that could form the basis for a longer answer to be done on your own paper.

ISBN: 9780170372855

Exam-type question 9

Enzymes are biological catalysts that enable plant and animal cells to carry out their metabolic functions. Scientists have proposed two models to explain how enzymes function.

Discuss enzyme function including:

- an explanation of both models. Use the box below to represent these.
- an explanation of how *three* different factors may affect enzyme activity.
- relevant named examples to illustrate your answer.

4

Use this scaffold framework as the basis for an answer to be done on your own paper.

- Start by defining the two main theories of enzyme function, making it clear how they differ.
- For example, 'The first model is called the … and involves …' (see page 66).
- 'The second model is the … It is so-called because …'
- Include terms/words like: enzyme, enzyme-substrate complex, substrate and product, shape and the specific names of the models.
- Choose/identify one factor that affects enzyme function.
- Describe how this factor affects enzyme function (e.g. increases/reduces rate of action).
- Explain the mechanism (processes) of how this factor affects enzyme function. Use diagrams if these will assist you.
- Use named examples from either plants or animals which show this effect.
- Choose/identify a second factor that affects enzyme function.
- Explain how this second factor can increase/decrease the rate of function.
- Again, choose named examples from either plants or animals which show this effect.
- Name a third factor that affects enzyme function.
- Explain how this third factor can increase/decrease the rate of function.
- Again, choose named examples from either plants or animals which show this effect.

ISBN: 9780170372855

Exam-type question 10

Cells in most organisms are restricted in size, as they divide repeatedly rather than just growing larger. Discuss this statement. Include the following in your answer:

- How the 'surface area : volume ratio' changes as a cell grows.
- How movement of named materials into and out of specific cells is affected.
- How this can relate to the position of organelles in a cell and their shape.

Use this page to create a key points plan or a mind map or scaffold that could form the basis for a longer answer to be done on your own paper.

ISBN: 9780170372855

Genetic variation and change

NCEA Achievement Standard 91157

Demonstrate understanding of genetic variation and change

2.5

Externally assessed, 4 credits.

Unit 1 | Sources of variation

For all living things that reproduce by sex, no two individuals are exactly the same. All humans are genetically unique, except for identical twins, triplets, etc. Genetic uniqueness is less common in plants and animals that reproduce non-sexually.

In some situations this uniqueness and variety is entirely due to **genes**, but many characteristics are also influenced by **environment**. Example: milk yield in cows depends partly on their breed (genetic influence) and partly on their food (environmental influence).

Biology 2.5 deals with ways in which genes influence characteristics. Variation refers to the way genes express themselves in 'finished product' phenotypes, including visible traits such as eye colour and tongue rolling. Any **discontinuous** characteristic is also known as a **trait**.

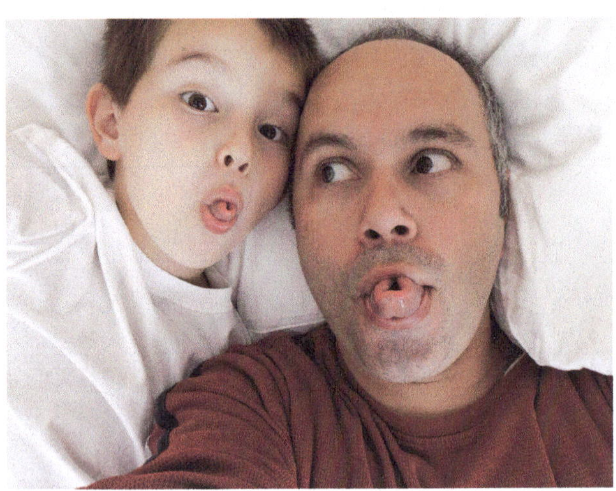

Tongue rolling, seen here in father and son, is a good example of discontinuous variation. You can either roll your tongue or not; there are no in-betweens. Height is an example of continuous variation. People are not either tall or short — there is a range of in-betweens.

Genetic variation is vital for evolution (see Unit 5). There are two causes of genetic variation — mutations and sex — with many subcategories:

- **mutations** (Unit 1; also Biology 2.7, Unit 3)
- **sexual reproduction**, in particular:
  - **meiosis**, including **chromosome segregation** and **crossing over** (Unit 2)
  - **fertilisation**, which introduces further variety (Units 2, 3, 4).

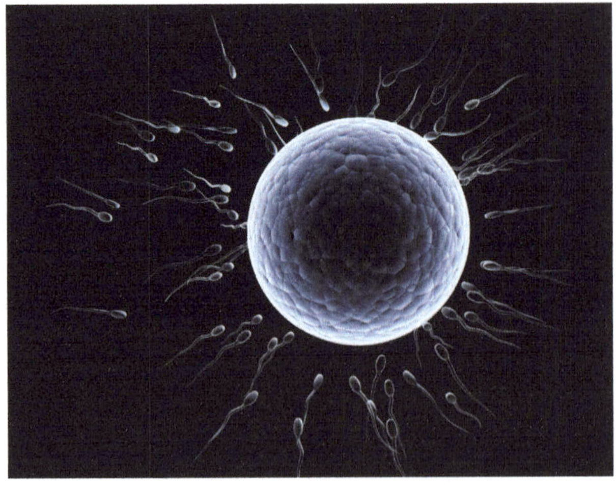

Fertilisation is a random event that increases genetic variation.

5

ISBN: 9780170372855

Chromosomes, genes, alleles

The three words above will be used frequently, so it helps to know what they mean.

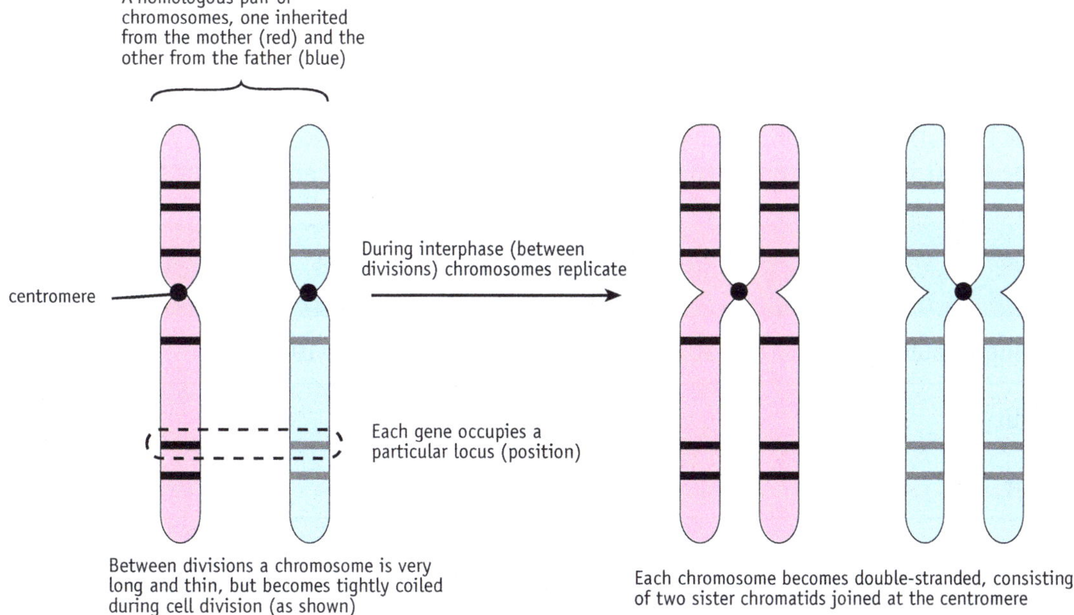

Fig. 2.5.1 A chromosome is a single molecule of DNA, containing thousands of genes (a few shown here as bands). Each gene occupies a distinct position called a locus (plural: loci).

All chromosomes exist in pairs, known as **homologous chromosomes**. Humans have 23 pairs, dogs 39 pairs, cats 19 pairs. One chromosome of each pair always comes from the father, one from the mother. A **gene** may be thousands of DNA base pairs long and there may be thousands of genes on one chromosome; for explanation of 'gene', see Biology 2.7, Unit 4. **Alleles** (pronounced *AL-eels*) are alternative forms of a gene. A pair of alleles in a body cell may be identical (the organism is **homozygous**), or they may be different (the organism is then **heterozygous**).

5

Activity A

1 Explain the difference between continuous and discontinuous variation, and give two examples of each.

2 Explain the difference between: gene, locus, allele.

ISBN: 9780170372855

3 Explain what is meant by a 'trait'.

4 Decide whether each of the following is influenced entirely by genes, or by environment, or by a mixture of each. Your answers will be a matter of opinion, but place scores totalling 10 in each vacant box. Example: a feature could score 7/10 for genetic influence, 3/10 for environmental influence.

Feature	Influenced by genes?	Influenced by environmental factors?
Tongue rolling in humans	10/10	0/10
Musical ability		
Adult height		
Eye colour		
Milk yield in cows		
Your accent		
Born with five toes on each foot		

5

Mutations

A **mutation** is a permanent change in DNA. Affected individuals are **mutants**. If a mutation is **gametic** (present in sex cells), it can be passed on to the next generation. Example: albinism. Any mutation originating in **somatic** (body) cells cannot be passed on, but some somatic mutations cause cancers.

There are two main types of mutation, with many subcategories of each:

- **gene mutations**, aka 'point' mutations, where the DNA base sequence of one gene is changed. Biology 2.7, Unit 3 has information on gene mutation causes, types, and effects.
- **chromosome mutations**, in which a whole section of a chromosome is changed, or else chromosome numbers are changed. Details below.

All types have the following important characteristics:

- Mutations are **rare**. Example: for achondroplastic dwarfism, the number of new cases is one per 71,000 births.
- Most mutations are **harmful** or else **neutral** ('silent').
- **Useful** mutations are even more rare.
- Evidence indicates that mutations are **random**, and also that useful mutations don't become more likely if they are needed.
- Most mutant genes are **recessive**, so are not expressed in heterozygotes.

ISBN: 9780170372855

E

Evolution

Natural selection is the main 'driver' of evolution, as described in Unit 5 and in greater detail in workbook 3. Genetic variation provides the raw material for natural selection to work on. Species with little variation — often the result of a genetic bottleneck — have less potential to evolve to meet changing circumstances. Large amounts of genetic variation within a species make evolutionary change more likely.

Sex reproduction increases genetic variation, because meiosis and fertilisation recombine (reshuffle) existing genes. Mutation can actually introduce new genes, but this is comparatively rare.

Whether a mutation or a meiotic change is useful or not depends on circumstances. Example: in polar bears white fur is very useful, but in most animals white colour harms their chances of survival.

Chromosome mutations

Changes in structure

Sections of a chromosome can be duplicated (replicated), inverted (turned around), translocated (swapped), or even completely missing. Fig. 2.5.2 illustrates what can actually happen. In any of these situations mutation can affect a number of genes with unpredictable results.

Extra or missing chromosome

An individual chromosome may be represented the wrong number of times. Examples:

- People with Down syndrome have 'trisomy 21', which means three versions of chromosome 21, instead of the normal two. Result: a total of 47 chromosomes instead of the normal 46.
- Women affected with Turner syndrome have only one X chromosome instead of the normal two, giving a total of 45 chromosomes.

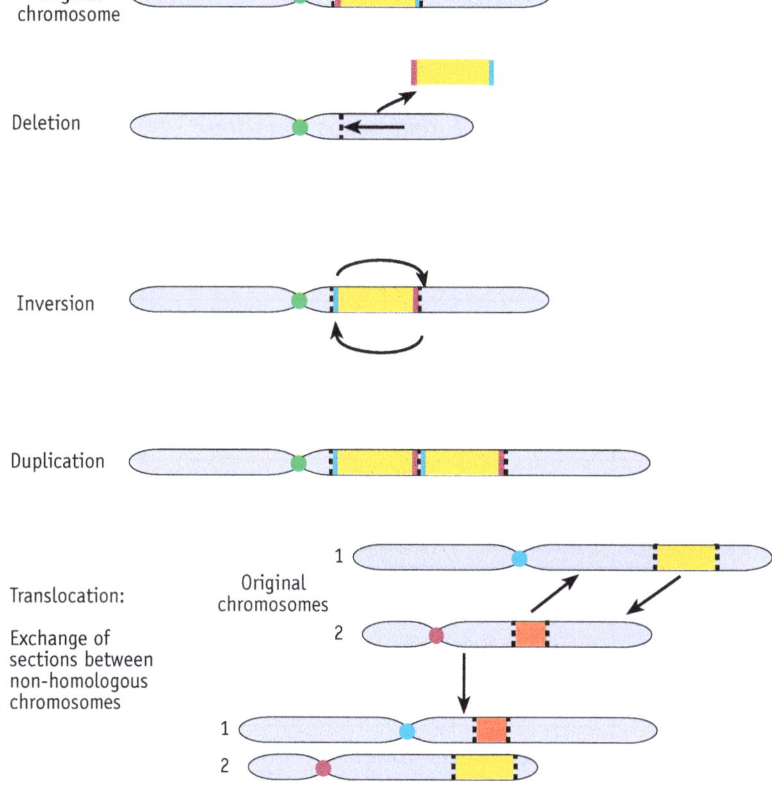

Fig. 2.5.2 Different kinds of chromosome mutation.

ISBN: 9780170372855

Polyploidy

In some situations an entire chromosome set may exist as multiple sets. Instead of the normal diploid number (2*n*), there may be three or more sets of chromosomes. Polyploidy is rare in animals but common in plants.

The normal number of chromosomes in each body cell varies from species to species. In humans it is 46. Chromosomes occur in pairs (in body cells), so the normal count is always an even number, given the general designation '2*n*'.

Type of cell	Normal number of chromosomes	Technical term for this number
Gametes (eggs, sperm)	*n*	haploid
Somatic (body) cells	2*n*	diploid
Polyploid mutant cells	3*n*	triploid
	4*n*	tetraploid
	6*n*	hexaploid

Table 2.5.1 Words and numbers associated with 'ploid' naming

Activity B

1 Write the word(s) that match each definition or description in the table below. Select from this list: *chromosome, locus, variation, mutation, polyploid, discontinuous variation, allele, point, expression, gene, trait, diploid, silent, continuous variation.*

a	Differences within one species	
b	A length of DNA responsible for one protein	
c	One form of a gene	
d	The position of a gene on a chromosome	
e	Characteristics with a whole range of in-betweens	
f	Characters that can exist in several distinct forms	
g	Ribbon-like structure containing DNA	
h	Any distinct inherited characteristic	
i	A genetic change	
j	The outward result of gene action	
k	A mutation that does not have any effect	
l	A mutation causing multiple sets of chromosomes	
m	A mutation in which a single DNA base is changed	
n	The normal number of chromosomes per cell	

2 The two main sources of genetic variation in most living things are ____________________

and ____________________

5

ISBN: 9780170372855

3 The letters A to J represent all the genes in one chromosome. In each mutation, identify the chromosome mutation as either deletion, duplication (repetition), inversion, or translocation.

normal: ABCDEFGHIJ

1st mutation: ABCDCDEFGHIJ ____________________

2nd mutation: ABCDEFJIHG ____________________

3rd mutation: ABCDEFGHIJKL ____________________

4th mutation: ABDEFGHIJ ____________________

4 Explain the difference between the kind of chromosome change that results in Down syndrome, and the kind of chromosome change that results in triploidy.

5 Suggest why useful mutations are even more rare than are harmful mutations.

6 Explain the main biological importance of genetic variation.

ISBN: 9780170372855

Unit 2 | Meiosis

Meiosis is a special kind of nuclear division that (in animals) results in the production of gametes (sex cells). Two key features of meiosis are:

- it **halves the number of chromosomes**
- it **ensures that all cells produced are genetically different** from each other. This uniqueness is a major source of genetic variation.

In animals, meiosis occurs only in the sex organs: ovaries and testes.

Sex and non-sex

Sex is not the only way of reproducing. Many plants and some animals can reproduce by asexual (non-sexual) methods. Some depend exclusively on asexual methods. Example 1: bacteria, which reproduce by binary fission. Example 2: banana trees, which have flowers but have not used them for reproduction for over 10,000 years, and are reproduced non-sexually by planting cuttings. In bananas and bacteria and others, mutation is the sole source of genetic variety. For them, meiosis is not an option.

Why half?

Why does the number of chromosomes need to be halved in meiosis?

Each **gamete** has ***n*** chromosomes — the **haploid** number,
so that after fertilisation
each **zygote** has **2*n*** chromosomes — the **diploid** number.

If gametes had 2*n* chromosomes, the number would double at every fertilisation. To avoid a chaotic situation, meiosis halves the chromosome number.

Gamete is the word for eggs and sperm. It is from the Greek word *gamos*, meaning marriage.
Zygote is the technical term for a fertilised egg.

Fertilisation

Fig. 2.5.3 represents cells in a sexually-reproducing species. In reality, kiwi have 80 chromosomes, but to keep things simple the drawing shows only four. Chromosomes inherited from the male parent are shown in blue, chromosomes from the female in red.

There is a simple way to avoid confusing the similar-looking words meiosis and mitosis. Meiosis has an E, as does sEx.

5

ISBN: 9780170372855

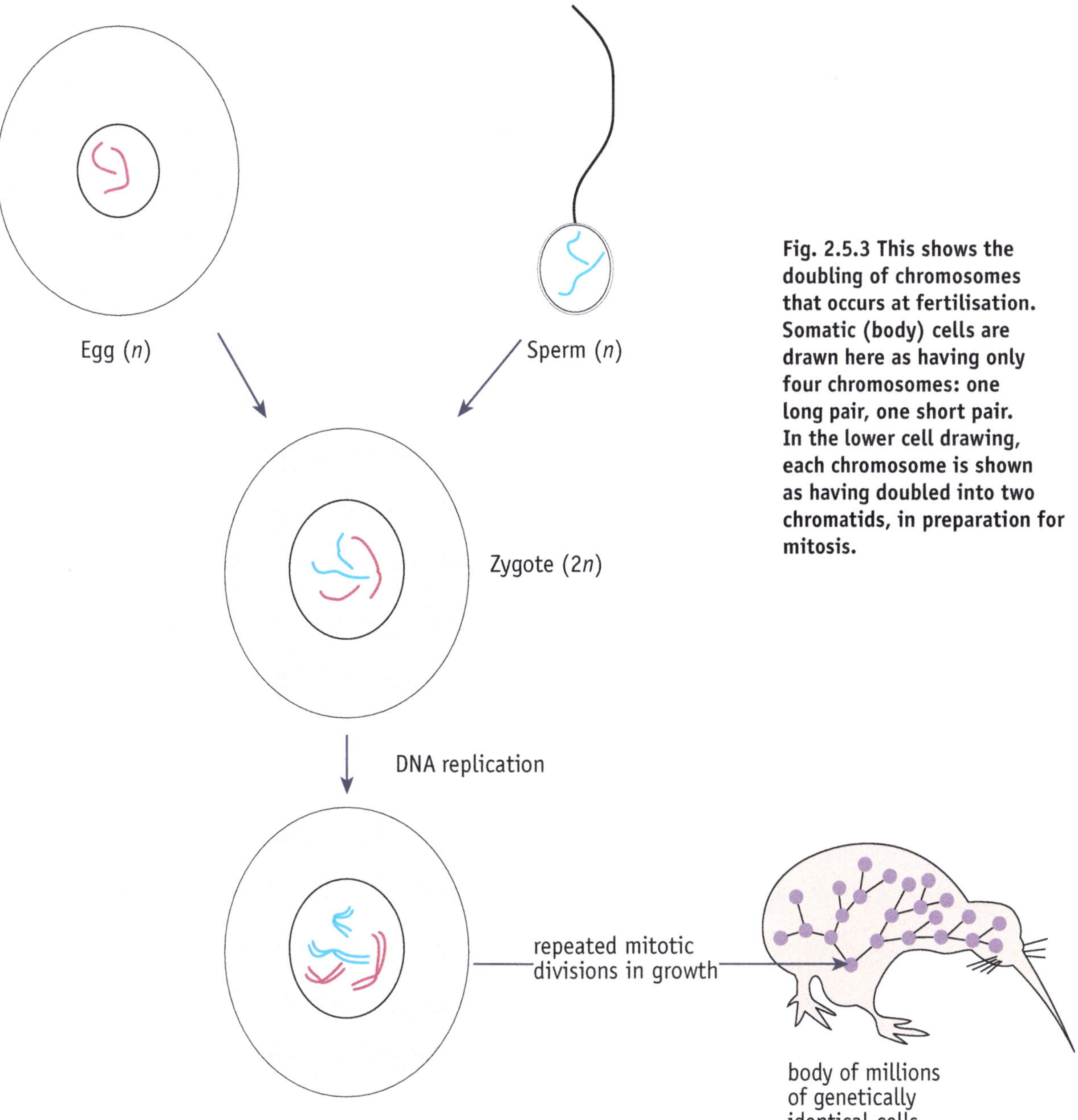

Fig. 2.5.3 This shows the doubling of chromosomes that occurs at fertilisation. Somatic (body) cells are drawn here as having only four chromosomes: one long pair, one short pair. In the lower cell drawing, each chromosome is shown as having doubled into two chromatids, in preparation for mitosis.

5

Meiosis details

The results of meiosis are more complicated than mitosis in three ways:

- meiosis has two nuclear divisions, producing four daughter cells
- meiosis reduces the chromosome number from $2n$ to n
- meiosis produces nuclei that are genetically different from each other — possibly unique.

It's important to understand the reason why the chromosome number is reduced to n, and also understand why genetic variation is an advantage. Fig. 2.5.4 explains how these changes are caused.

ISBN: 9780170372855

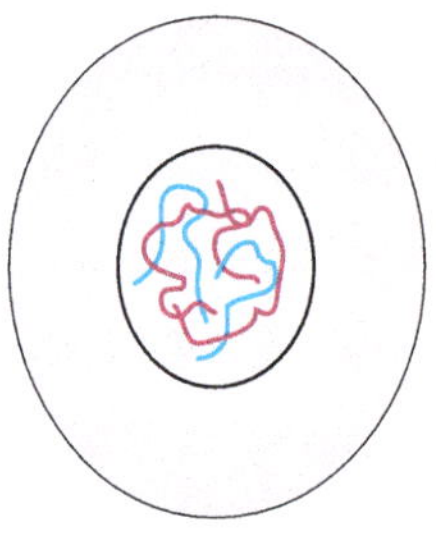

1. Before meiosis, chromosomes are single-stranded and invisible (but shown here for convenience)

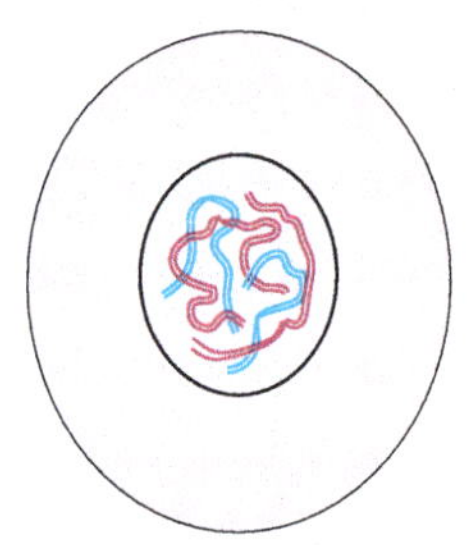

2. Chromosomes replicate, each now consists of two identical chromatids

MEIOSIS I

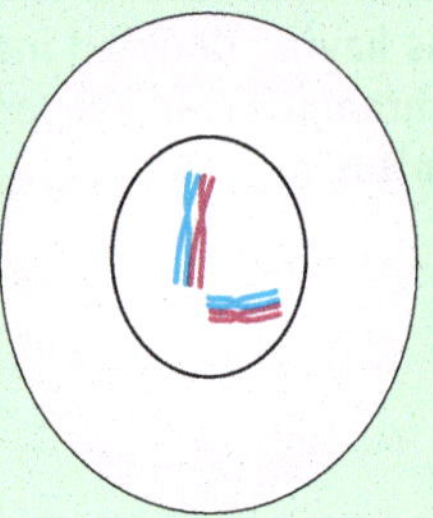

3. Homologous chromosomes come together in pairs

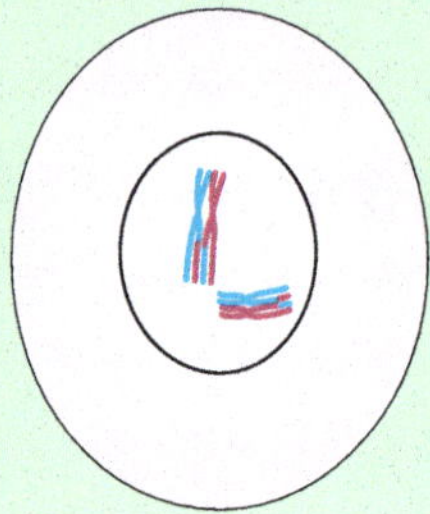

4. Crossing over occurs; 'non-sister' chromosomes exchange sections

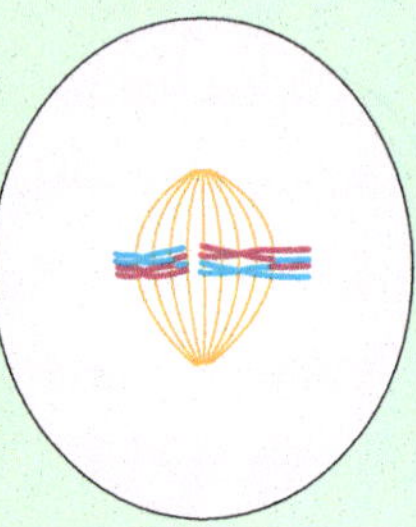

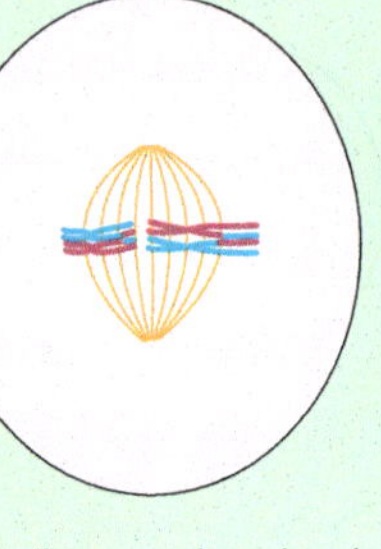

5. Nuclear envelope breaks down and spindle forms. Centromeres attach to spindle at its equator

Fig. 2.5.4 Important events associated with meiosis are shown in green boxes. There are two nuclear divisions: meiosis I and meiosis II. Events that occur just before and after meiosis I and II are shown in boxes framed by dotted lines.

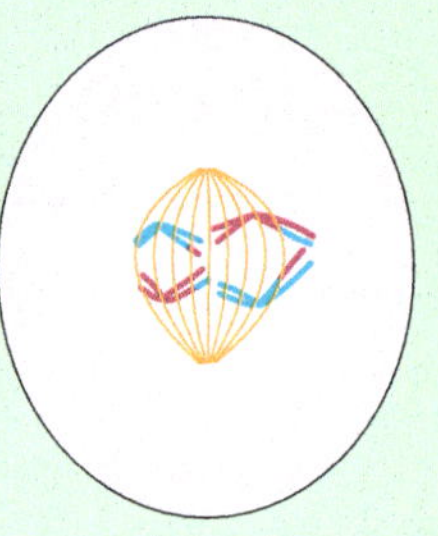

6. Homologous chromosomes dragged to opposite poles of the spindle

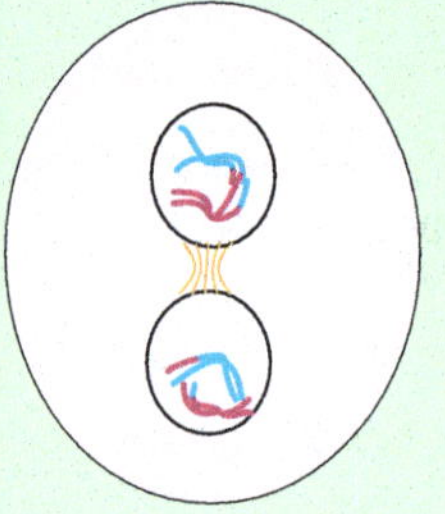

7. Each pole of the spindle receives a haploid set of double-stranded chromosomes

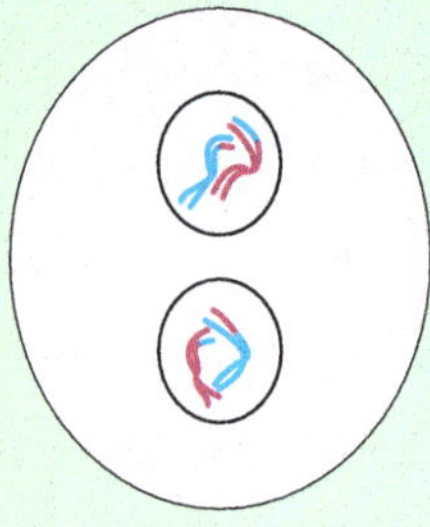

8. A nuclear envelope develops round each chromosome cluster

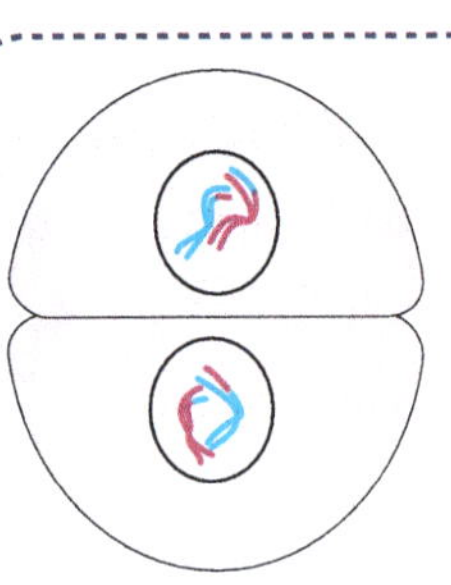

9. Cytoplasm divides to form two genetically different cells

MEIOSIS II

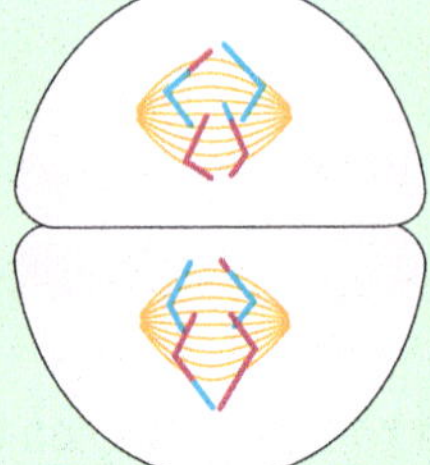

10. Chromatids are dragged to opposite poles of spindle

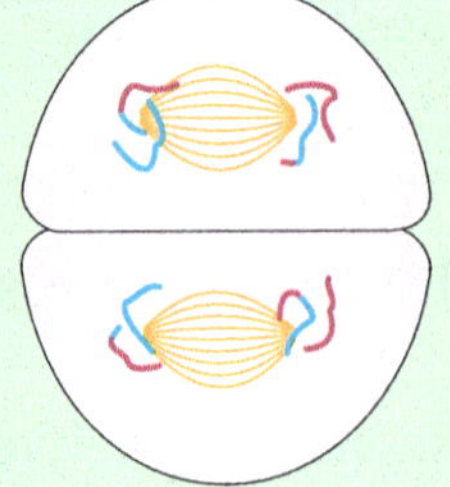

11. Each pole of the spindle receives a haploid set of single-stranded chromosomes

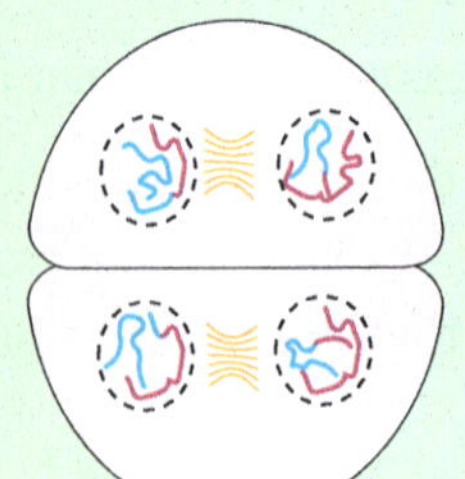

12. Chromosomes become indistinct; nuclear envelope develops round each genetically different haploid set

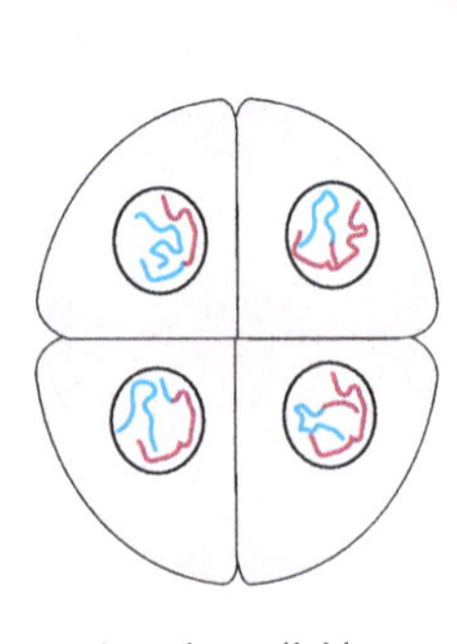

13. Cytoplasm divides to form four genetically different cells

ISBN: 9780170372855

Crossing over

Before meiosis begins, the chromosomes are replicated (copied). Each now consists of two identical **chromatids**, joined at a **centromere**. Early in meiosis I (stages 4 and 5 in Fig. 2.5.4), chromosomes come together as **homologous pairs**. At points exactly opposite each other, 'non-sister' chromatids belonging to different members of each chromosome pair break. ('Non-sister' refers to chromatids from different homologous chromosomes. It has nothing to do with sisters.) The point of breakage is the **chiasma**. 'Non-sister' ends then join the 'wrong' chromatid. The affected chromatids now consist of mixed maternal and paternal segments. This break-and-rejoin process is called **crossing over** and it reshuffles genes into many new combinations, de-linking alleles that were originally on the same chromosome.

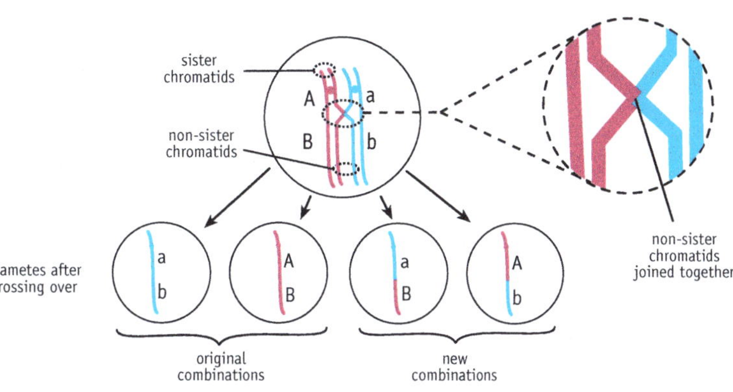

Fig. 2.5.5 How crossing over creates new gene combinations. The drawing shows four new combinations. The letters represent alleles. Gametes *ab* and *AB* are different as a result of independent assortment. Gametes *aB* and *Ab* are different as a result of crossing over.

Independent assortment

In meiosis I, homologous chromosomes are dragged to opposite ends of the spindle, as shown in Fig. 2.5.4, stages 5, 6, 7. It is a matter of pure chance to which end a particular chromosome moves. They all move independently of each other. Segregation of different chromosome pairs causes **independent assortment**, which 'reshuffles' genes located on different chromosomes into new combinations. This happens whether or not there has also been crossing over. Also see Unit 4.

Massive numbers

There are many chromosome pairs (23 pairs in humans), and thousands of genes on each chromosome. Crossing over can happen at any point between these many genes. Also, we know that double and triple crossing over can occur, with several bits of chromosome being exchanged between homologous chromosomes. In addition, it is a matter of pure chance which one of millions of sperm will get to fertilise any one egg. When all these factors are taken into account, the number of new possibilities in just one fertilisation is perhaps greater than 10^{100}, more than the number of atoms in the known universe. The whole system favours uniqueness, and explains why siblings are never identical — except for identical twins, triplets, etc.

Activity A

1 Write the word(s) that match each definition or description in the table below. Select from this list: *diploid, meiosis, somatic, gamete, fertilisation, mitosis, crossing over, centromere, haploid, zygote.*

a	General word describing eggs and sperm	
b	A fertilised egg	
c	Process of sperm chromosomes entering egg	
d	Full set of chromosomes in body cells	
e	Number of chromosomes in a gamete	
f	A technical word referring to body cells	
g	Causes new gene combinations in a chromosome pair	
h	Central part of a chromatid attaching to spindle	
i	Cell division that results in gamete production	
j	Normal cell division, producing somatic cells	

2 Complete the following sentences.

a In meiosis, the number of chromosomes is halved because

b Bacteria rely on mutations to create variety because

c The difference between centromere and chiasma is

d The biological importance of independent assortment and crossing over is

 ISBN: 9780170372855

3 Complete this table to summarise differences between mitosis and meiosis in animals.

	Mitosis	Meiosis
Occurs where (in animals)?		
Resulting cells haploid or diploid?		
Resulting cells genetically identical?		
Chromosome number reduced or unchanged?		

4 a Describe one way in which meiosis I resembles mitosis.

b Describe one way in which meiosis I differs from mitosis.

5 In humans there are 46 chromosomes in each body cell.

a How many pairs of chromosomes is this? ______

b What is the human diploid number? ______

c What is the haploid number? ______

d How many chromosomes in an egg nucleus? ______

e How many chromatids per nucleus at an early stage of meiosis I? ______

6 Banana tree plantations are clones — lots of genetically identical individuals.

a Explain why this increases the likelihood of fungus diseases causing widespread damage.

b Suggest how genetic variation is occasionally created in bananas.

ISBN: 9780170372855

Symbols and genotypes

It's not practical to keep drawing pictures of chromosomes and alleles, so the custom is to represent alleles with letters. There is no strict rule as to which letter of the alphabet should be used, but mostly we use capital letters for dominant alleles, and lower case letters for recessive alleles.

Table 2.5.2 Examples of symbol use

Trait	Symbol for dominant allele	Symbol for recessive allele
Eye colour, brown or blue*	*B*	*b*
Able to roll tongue, or not	*R*	*r*
Blood groups**	I^A	I^O (also written i^O)

*** In reality, several pairs of alleles are involved in eye colour.**
**** In some cases, superscripts are used for allele symbols.**

Genotype is an individual's gene makeup. In body cells, chromosomes always occur in pairs — which means that alleles also occur in pairs. For eye colour alone, the genotype could be written as:

BB or *Bb* or *bb*

If we use symbols to represent both '*B*' and '*R*' genes mentioned in Table 2.5.2, then we can show the nine possible genotypes like this:

BBRR, BBRr, BBrr, BbRR, BbRr, Bbrr, bbRR, bbRr, bbrr

Let's look at just one of these genotypes: *BbRr*. Assume that the eye-colour and tongue-roll genes are linked, with *BR* on one chromosome and *br* on the 'partner' chromosome, and that there is no crossing over. This means that there are only two different possible combinations in gametes:

BR and *br*

Let's look again at the *BbRr* genotype. Assume that the eye-colour and tongue-roll genes occur on different chromosomes, and are allocated ('sorted') independently when chromosomes are separated during meiosis. This time there are four different possible combinations in the gametes:

BR, Br, bR, br

Remember: because of meiosis, each gamete carries only one allele for each character, never two.

ISBN: 9780170372855

Activity B

1 The drawing represents a cell with one pair of chromosomes, drawn at an early stage of meiosis I, with four chromatids and crossing over about to happen. Draw red and blue chromatids in the six blank cells to show the resulting arrangement at different stages.

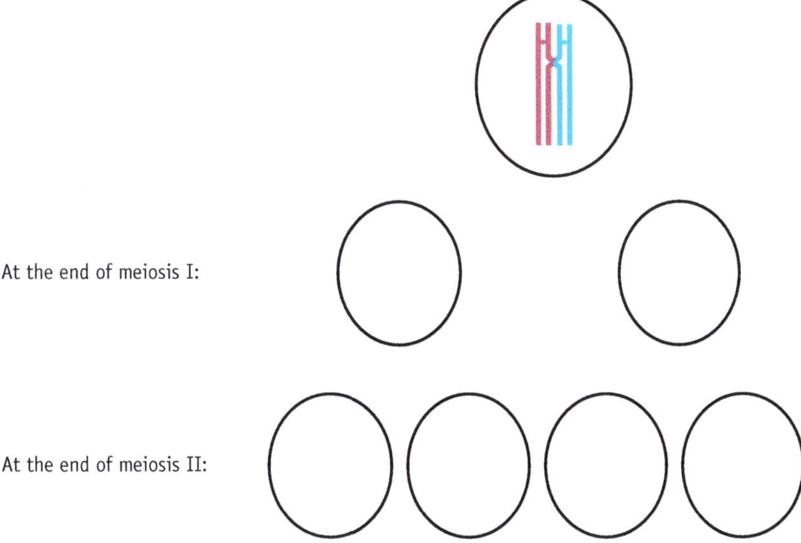

2 Genes *A* and *D* are linked together on the same chromosome. If *no* crossing over happens during meiosis, predict the possible gametes that could be produced by each of these genotypes:

a *AADD* ____________________

b *AaDd* ____________________

c *Aadd* ____________________

3 Genes *A* and *D* are linked together on the same chromosome. If crossing over *does* happen, predict the possible gametes that could be produced by each of these genotypes:

a *AADD* ____________________

b *AaDd* ____________________

c *aadd* ____________________

4 Genes *A* and *K* and *E* are all on different chromosomes. Predict the possible gametes that could be produced by each of these genotypes:

a *AAKKEE* ____________________

b *AaKkEe* ____________________

c *AAKkEe* ____________________

5

ISBN: 9780170372855

Unit 3 | Monohybrid inheritance

Remember these five basics

1. Chromosomes occur in **homologous pairs**, one inherited from each parent.
2. Most genes can exist in two or more alternative forms, known as **alleles**. Each allele occupies the same **locus** (position) on the two homologous chromosomes.
3. Because of meiosis, **each gamete carries only one allele** for each character, never both.
4. In many cases, one allele is **dominant** and the other is **recessive**. Dominant alleles are usually represented by upper-case letters, and recessive by lower case.
5. If an organism has two of the same allele, it is **homozygous** (examples: *AA, aa*). If it has two different alleles, it is **heterozygous** (example: *Aa*). Recessive alleles can only show a phenotype effect if they are in homozygotes. (Note that 'homozygous' and 'heterozygous' are adjectives; 'homozygote' and 'heterozygote' are nouns.)

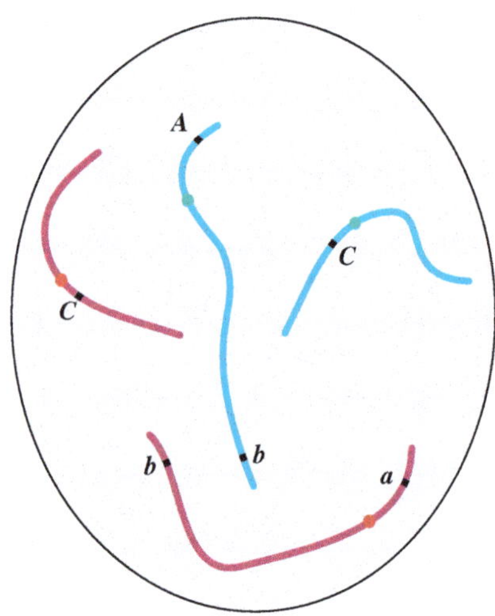

Fig. 2.5.6 An arrangement of three genes on two chromosome pairs. Each gene occupies a particular locus, which becomes more obvious when homologous chromosomes are placed alongside each other. This imaginary cell (and organism) is heterozygous for gene *Aa*, homozygous recessive for *bb*, homozygous dominant for *CC*. Overall genotype: *AabbCC*.

Genotype: an individual's inherited genetic information.
Phenotype: the physical results of gene action as expressed ('shown') in the body.

Compare this to a plan and a house. A plan represents the genotype and the finished product, the house, is the phenotype.

Solving genetics problems

We can't see genes, but we can study the way they are inherited by looking at a family tree, or else by doing breeding experiments (in plants and animals but not people!). What kind of experiments?

- The most useful traits are those that have clear-cut groups. For example, pea plants are either short or tall, and have purple or white flowers. These examples of **discontinuous variation** are little affected by differences in the environment.
- The simplest kind of breeding experiment is a **monohybrid cross**. This involves one gene with two alleles.

 ISBN: 9780170372855

Mouse monohybrid cross

These drawings show the results of two crosses between mice, in both cases **monohybrid**, which means they involve only one gene. All the mice are either black or brown. The parents are both **true-breeding**, which means both come from lines that have bred only that colour for many generations — so must be homozygotes.

The offspring of two contrasting homozygotes are the **F_1 generation**. If these heterozygous F_1 individuals grow up and are allowed to breed among each other, the result is the **F_2 generation**.

The lower drawing (Fig. 2.5.8) shows how chromosome and allele behaviour can explain the results. For any such explanation, it is important to start with correct genotype for the parents. The gamete genotypes are then written outside the **Punnett square**. Inside the boxes we write genotypes that would result from each fertilisation.

A Punnett square is used to show the different ways in which different kinds of gamete can join in fertilisation. In this example there are four ways that gametes can combine, resulting in three different genotypes and two different phenotypes.

Finally, underneath the Punnett square we write numbers to indicate the **expected proportions** of young that are likely to be produced after many repeat trials. These proportions (results) can be written in any of four ways, as:

- fractions
- percentages
- phenotype ratios
- genotype ratios

In this case, the F_2 includes 25% brown mice. Their F_1 parents were both black, but carried recessive alleles for brown.

In this example, eggs have been written on top, sperm at the side. A Punnett square works just as well if eggs are written at the side.

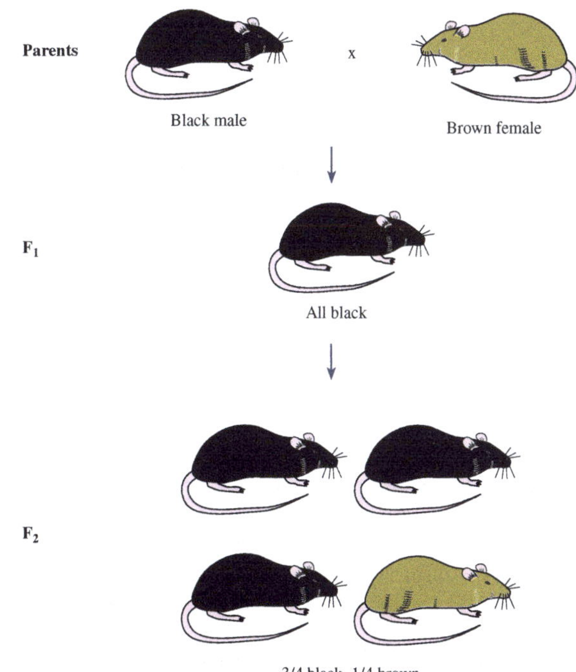

Fig. 2.5.7 The results of a cross between a brown mouse and a black mouse.

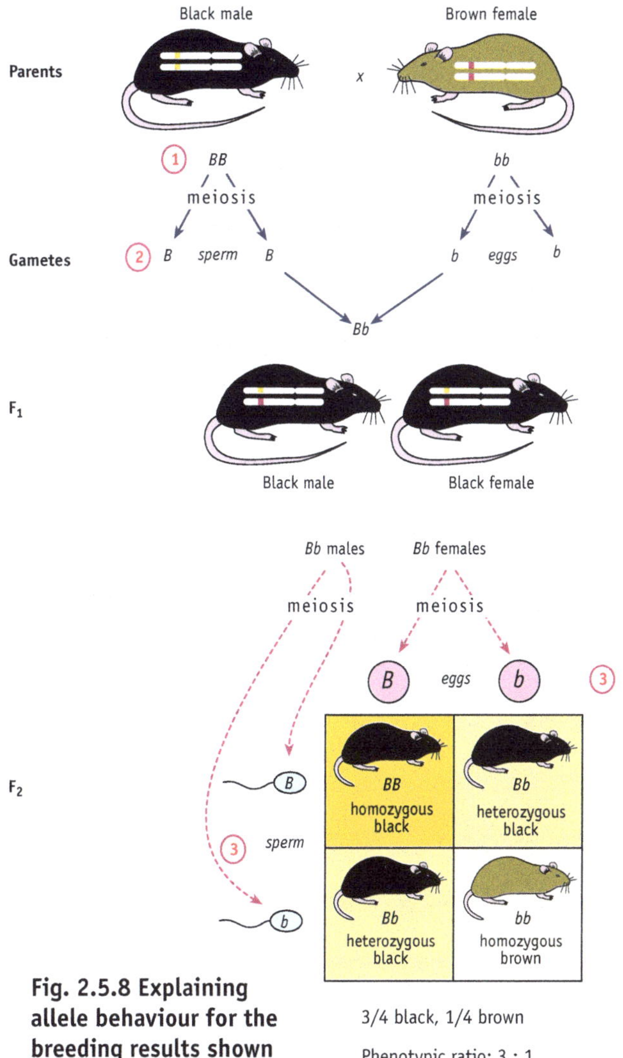

Fig. 2.5.8 Explaining allele behaviour for the breeding results shown in Fig. 2.5.7.

ISBN: 9780170372855

E

Predicted and actual numbers are usually different

Punnett squares are used to predict ratios. Example: a human couple with brown eyes (but both parents are heterozygous) could be predicted to have a 3:1 ratio of brown-eyed and blue-eyed children, i.e. 25% blue-eyed children. However, these are unlikely to be actual numbers, for two reasons. First: if the couple have two or three children, we can't say that 25% will be blue-eyed. Second, any predictions are probabilities, not certainties. Even if the couple do have four children, the actual 'brown-to-blue' ratio could end up 2:2, or 3:1, or 1:3, or 4:0. Punnett squares give predicted ratios that are likely to show up with large numbers of offspring, not actual numbers when there are few offspring. Each fertilisation is a random event not influenced by previous fertilisations.

Testcrosses

Testcross: a method that is used to find whether an individual is homozygous or heterozygous.

In farm animals and plants there are many situations where a breeder needs to develop a pure line, all with the same alleles. This is not easy because homozygous and heterozygous individuals can look the same. (Example: black mice can be *BB* or *Bb*.)

A testcross is done by breeding the individual in question with an individual with the recessive phenotype. Fig. 2.5.9 shows two possible results, using mice as an example.

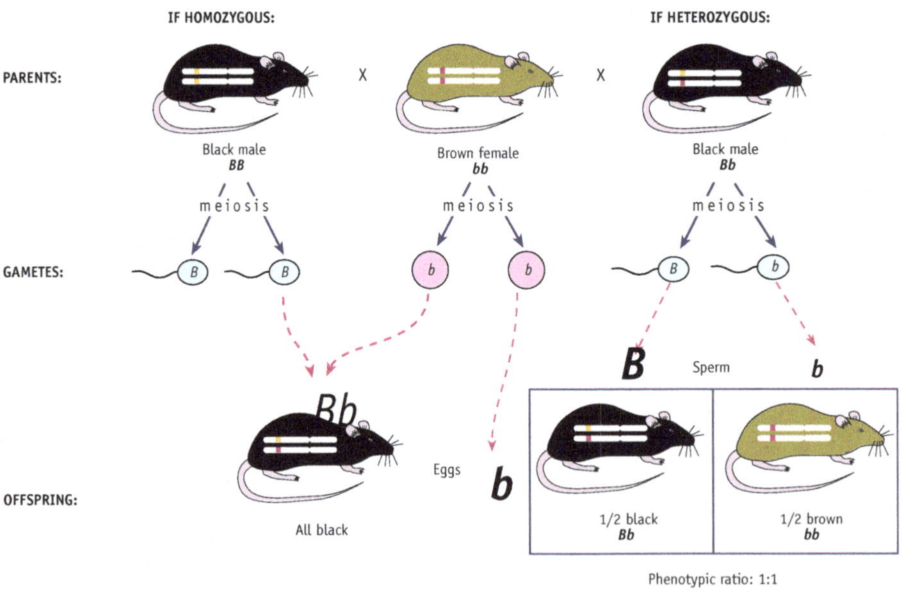

Fig. 2.5.9 An example of a monohybrid testcross. If all offspring are black, this proves the black parent is homozygous (*BB*). If some offspring turn out black and some brown, this proves the black parent is heterozygous.

 ISBN: 9780170372855

Activity A

1 Write the word(s) that match each definition or description in the table below. Select from this list: *locus, recessive, genotype, true breeding, discontinuous, phenotype, monohybrid, alleles, F_1, testcross, F_2, dominant.*

a	An individual's inherited genetic information	
b	Characteristics that are expressed by gene action	
c	An allele that is expressed only in homozygotes	
d	Alternative forms of a gene	
e	An allele that is expressed even in heterozygotes	
f	Controlled by only one pair of genes	
g	First-generation offspring from two pure lines	
h	Offspring resulting from breeding F_1 individuals	
i	Have produced only that trait for many generations	
j	Used to detect heterozygous individuals	
k	Variation that occurs with no in-betweens	
l	The position of a gene on a chromosome	

5

2 A brown mouse is mated with another brown mouse. State the genotype of each mouse. (Use the symbols provided in Fig. 2.5.8.) ____________________ Can they produce black offspring? (Yes/No) ________ Justify your answer.

____

____

____

3 A female brown mouse is pregnant, and is known to have mated with a black mouse whose genotype is uncertain. The female mouse eventually gives birth to three brown and five black babies.

a State the genotype of the father (as regards colour). ____________

b Justify your answer.

____

____

____

ISBN: 9780170372855

c State the genotype of the father's gametes. ____________

d Complete a Punnett square to show possible genotypes resulting from this particular mating. Choose whichever of the two versions here is more suitable for the purpose.

e Predict what fraction of brown babies you would expect from this mating. ____________

f State what fraction of brown babies were actually born.

____

g Suggest why the above two fractions are slightly different.

____

____

____

4 Explain the purpose of a testcross.

____

____

____

5 State which phenotype is always used in any testcross.

____

Remember: because of meiosis, each gamete carries only one allele for each character, never both alleles.

6 Refer to Fig. 2.5.8. This mating between black mice produced eight babies. According to predictions we would expect two of these eight to be brown. In fact only one of the eight was brown. Explain why.

____

____

____

____

____

____

____

____

 ISBN: 9780170372855

Incomplete dominance

Incomplete dominance is a situation where the heterozygote's phenotype is intermediate between the two homozygote phenotypes. Example: if a purebred chestnut horse is mated with a purebred white horse, the foals are all palomino — with a yellowish coat. Because neither allele is dominant, upper-case letters can be used for both. In this horse example, *C* is used for the chestnut allele, and *W* for the white. Both are on the same locus.

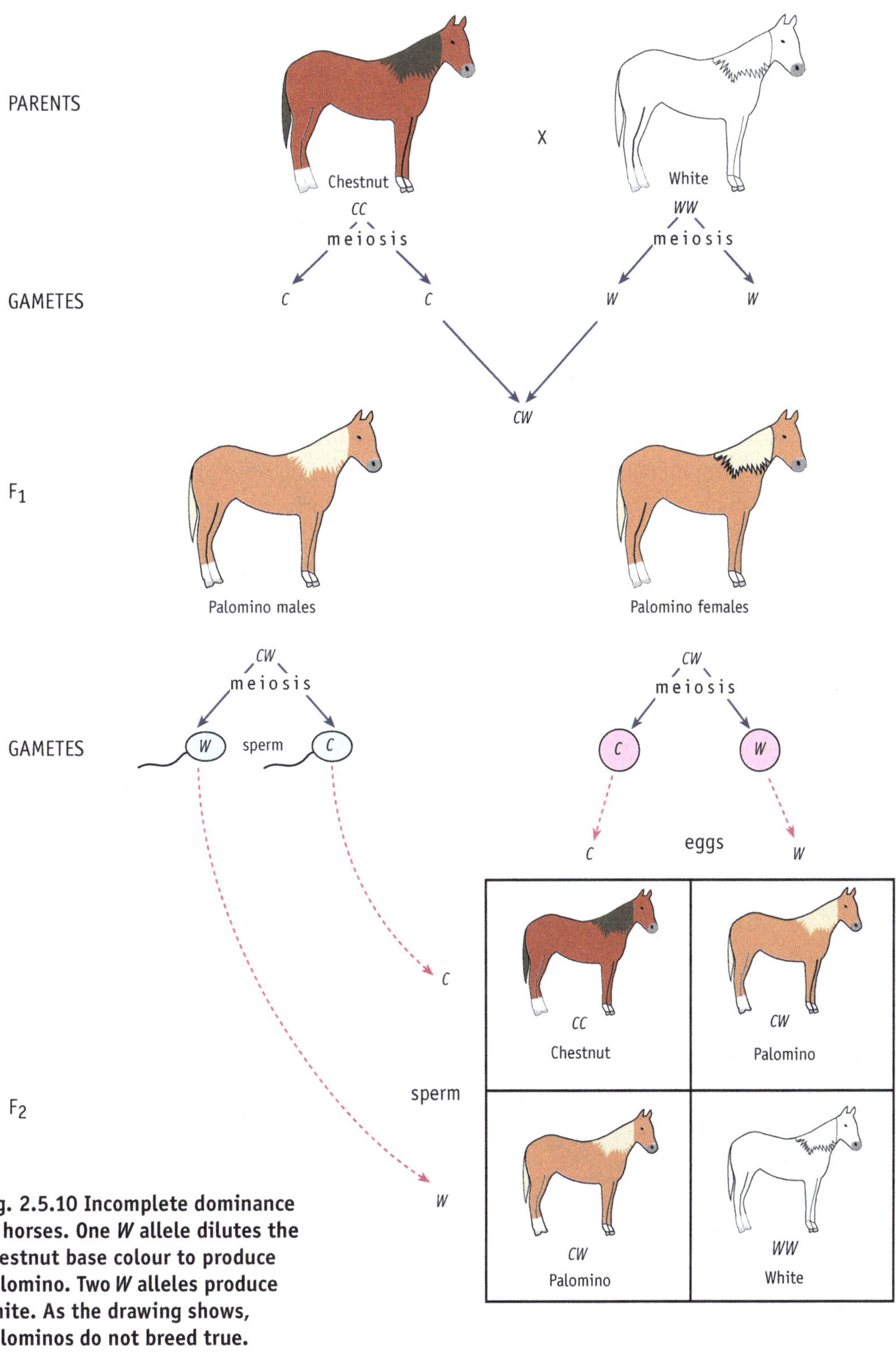

Fig. 2.5.10 Incomplete dominance in horses. One *W* allele dilutes the chestnut base colour to produce palomino. Two *W* alleles produce white. As the drawing shows, palominos do not breed true.

ISBN: 9780170372855

Multiple alleles and co-dominance

Not all genes exist in only two allelic forms; some have **multiple alleles**. Example: the ABO blood groups, which are controlled by three alleles, I^A, I^B and I^O (also written i^O). This gives four different phenotypes (Table 2.5.3). Every human's blood is group either A, B, AB or O — and these groups occur in all parts of the world. (There are other blood groups in addition to these, such as Rh+ and Rh–.)

Phenotype	Possible genotypes
Blood group A	I^AI^A or I^AI^O
Blood group B	I^BI^B or I^BI^O
Blood group AB	I^AI^B only
Blood group O	I^OI^O only (also written i^Oi^O)

Table 2.5.3 Blood group phenotypes, and the allele combinations that cause them. Note: There are three alleles in most human populations, but each individual carries only two of these.

Note: We use superscript allele symbols like I^A because if we used plain *A* it could cause confusion with phenotype A. I^A and I^B are co-dominant — see below. Both of these alleles are dominant over I^O.

Co-dominance is a situation where two alleles are expressed equally. This situation is quite common, and 'simple dominance' is less common. The inheritance of blood groups involves multiple alleles, and also provides an example of co-dominance. For someone with blood group AB, both the I^A and the I^B allele are expressed, and blood cells carry both kinds of protein on their surface: A and B.

Incomplete dominance

In snapdragons, red flower colour (*RR*) is incompletely dominant over white (*rr*), with the heterozygote (*Rr*) being pink. When the genes are expressed, *Rr* flowers produce two proteins, one that produces red pigment and another that produces no pigment, so *Rr* gives half as much red pigment as in *RR* flowers (Fig. 2.5.11).

Snapdragons.

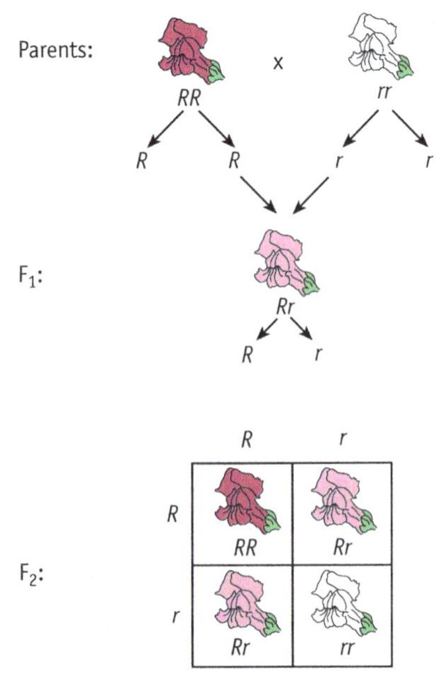

Phenotypic ratio: 1/4 red : 1/2 pink : 1/4 white

Genotypic ratio: 1 *RR* : 2 *Rr* : 1 *rr*

Fig. 2.5.11 Incomplete dominance in snapdragons.

ISBN: 9780170372855

Lethal alleles

Lethal genes were discovered when someone investigated the inheritance of coat colour in yellow mice. When a heterozygous x heterozygous cross is done, the phenotype ratio expected in the next generation should theoretically be three yellow to one grey. But the observed ratio for yellow mice is in fact always about 2:1. Why? Test crosses showed that all yellow mice are heterozygotes (*Yy*), and that the *Y* allele for yellow colour is dominant over *y* for grey. Further tests showed that all homozygous *YY* embryos die at an early stage, so we describe the *Y* allele as 'lethal'. Another example, Manx cats, page 146.

The gene for achondroplastic dwarfism provides a human example of a lethal allele. Affected individuals are all heterozygous. Having two alleles for this condition is lethal, resulting in the death of any fertilised egg that happens to inherit two of these alleles from two dwarf parents.

Activity B

1 Joshua has blood group O, Celia blood group AB, and their first baby is due.

a Joshua's genotype for his blood group is ______________________

b All his gametes will carry ______________________

c Celia's genotype for her blood group is ______________________

d Celia's gametes could carry ______________ or ______________

e Complete a Punnett square to show possible blood groups of their baby. Choose whichever of the two versions below is more suitable for the purpose, and state why the other version is less suitable.

____

____

f Explain why it is not possible for them to have a group AB child.

____

____

2 Megan and Daniel both have blood group A. Their first baby is group O.

a The genotype for blood group A could be ______________ or else ______________

b Which of these two possibilities is reality in their case? ______________________

ISBN: 9780170372855

c Explain how you decided that Megan and Daniel had this particular genotype.

d Complete a Punnett square to show all possible blood groups for this couple's children. Choose whichever of the two versions below is suitable for the purpose.

e Expressed as percentages, state the chances of Megan and Daniel having a first child who happened to be group A.

f Explain why it is not possible for them to have a group AB child.

3 Explain the difference between co-dominance and incomplete dominance.

4 Name a single example in each case of:

a a lethal allele

b multiple alleles

c co-dominance

d incomplete dominance

5 Explain why palomino horses do not breed true, and why about half their foals are not palomino. Use a Punnett square and base your explanation on information in Fig. 2.5.10.

 ISBN: 9780170372855

6 Snapdragon flowers can be red or pink or white. Complete the Punnett square below, which represents a cross between a pink-flower plant and a white one. Let *F* represent the allele for red flowers, *f* the allele for white flowers.

The phenotype percentages predicted from this cross will be red: ______ white: ______ pink.

7 Use the information on page 117 to find the possible outcomes of a mating between two heterozygous yellow mice.

a Parent genotypes: ______________________ and ______________________

Punnett square to show possible outcomes of this mating:

b Expected phenotype ratio resulting from this mating: ______ yellow : ______ grey

c Explain how this ratio occurs.

ISBN: 9780170372855

Sex-linkage

In all mammals, two of the 46 chromosomes determine (decide) sex. These **sex chromosomes** are known as X and Y. Females are XX, males are XY. The non-sex chromosomes are known as **autosomes.** (Autosomes each carry thousands of genes but autosomes do not 'decide' sex, so are not included in the explanations here.)

The Y chromosome is smaller than the X, and carries far fewer genes. There are many gene loci on the X that are absent on the Y. Genes that are on the non-matching part of the X chromosome are said to be **sex linked.** Because these genes are located on the X, we generally use allele symbols such as X^b and X^B.

Sex linkage example 1: **haemophilia.** Example 2: red-green **colour-blindness**, a harmless condition that occurs in roughly 8% of men and 1% of women (Fig. 2.5.12).

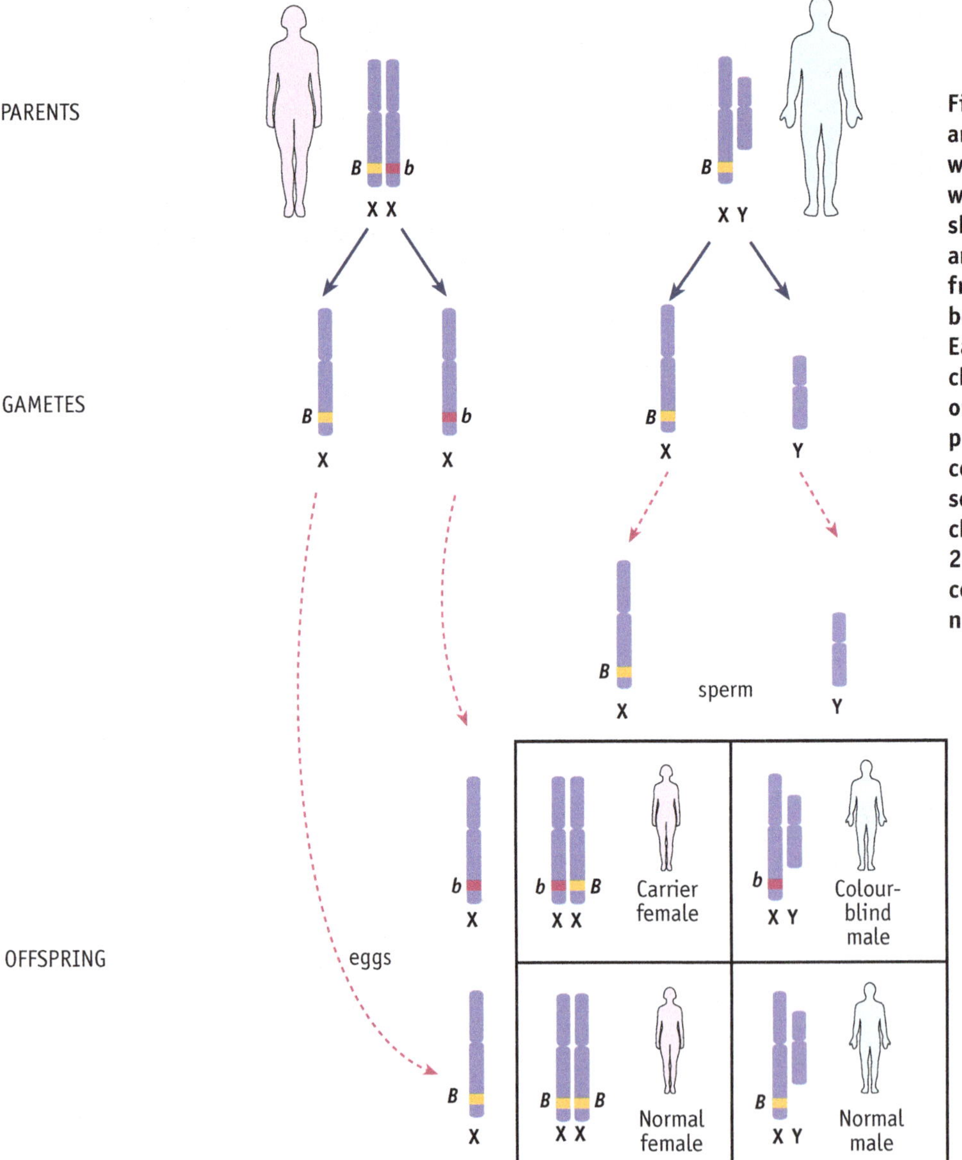

Fig. 2.5.12 The parents are a normal man, and a woman whose own father was colour blind — so she must have received an X^b chromosome from him, so she must be heterozygous. Each of this couple's children could have any one of four different phenotypes as regards colour-blindness and sex. Predictions: 25% chance of normal female, 25% carrier female, 25% colour-blind male, 25% normal male.

The genes for haemophilia and colour-blindness are linked on the X chromosome. Any genetics explanation that applies to one applies equally to the other.

5

ISBN: 9780170372855

Family trees

A family tree is a diagram that shows the phenotypes of related people, and the links between them over several generations. (A pedigree is similar to a family tree but does not show siblings.) Males are represented by squares and females by circles. Mutant traits are usually shown in a dark colour. Family trees often reveal whether a particular allele is dominant or recessive, autosomal or sex-linked. Four examples are shown below.

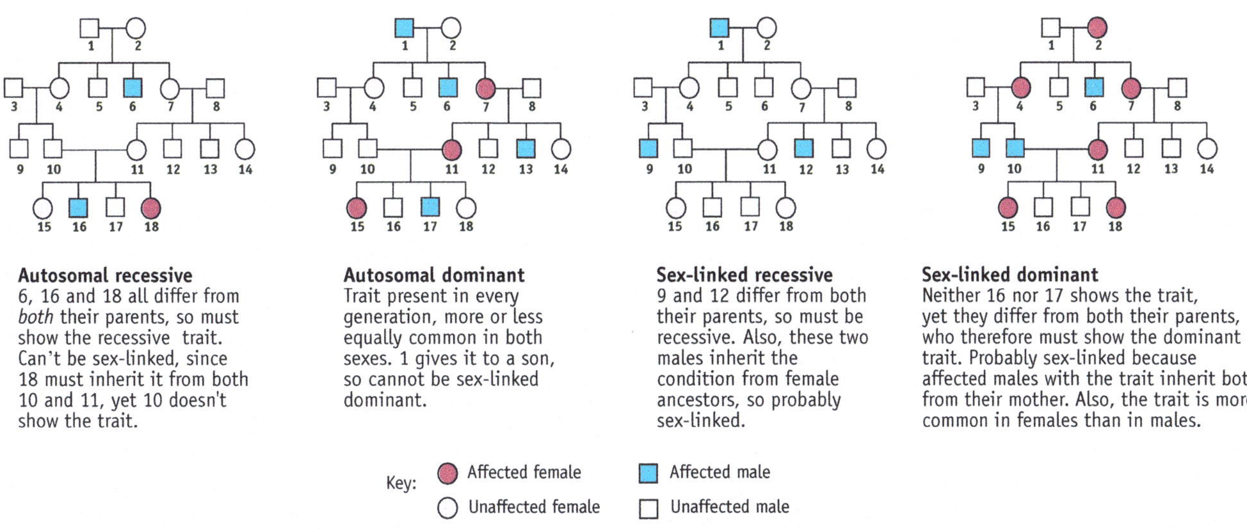

Fig. 2.5.13 Four different family trees, each illustrating a different pattern of inheritance.

Activity C

1 Explain what is meant by 'sex linked'.

2 Name two examples of traits that are sex linked.

3 This family tree represents three generations with 10 individuals, two of them colour-blind and shown here in black. Refer to this tree when answering **a** to **h**. Use the symbol X^E for the normal dominant allele and X^e for the recessive allele.

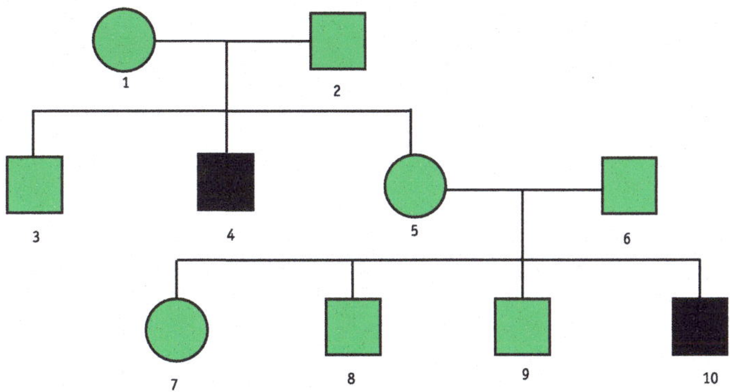

ISBN: 9780170372855

a Colour blindness is sex linked. Identify features of this family tree that provide supporting evidence for this fact.

____

____

____

b For grandparents 1 and 2 there had been no history of colour-blindness in either of their families over many previous generations. Give a genetic explanation of how colour-blindness appeared in this family.

____

____

c Identify two individuals who are not colour-blind, but are definitely carriers of the allele.

____

d Identify one other individual who is possibly a carrier of the allele.

____

e Complete a Punnett square to show the gametes and possible fertilisations in the marriage between 5 and 6. Choose the version that you consider suitable.

f In the marriage between 5 and 6, give the predicted likely numbers of their children's phenotypes if they have four children.

normal female ____________

colour-blind female ____________

carrier female ____________

normal male ____________

colour-blind male ____________

 ISBN: 9780170372855

g State the actual numbers of these phenotypes in the children of 5 and 6.

normal female ______________

colour blind female ________________

carrier female ______________

normal male ________________

colour blind male __________________

h Suggest reasons for the difference between predicted and actual proportions of their children.

____

____

____

____

4 State what genotype parents could possibly have a colour-blind daughter.

____

____

5 Explain why colour-blindness is much less common in females than in males.

____

____

____

6 Refer to Fig. 2.5.13 on page 121. For the family tree on the left, summarise the evidence that suggests this trait is:

a Recessive. ____

____

b Autosomal. ____

____

ISBN: 9780170372855

Unit 4 | Dihybrid inheritance

Dihybrid crosses with no linkage

The previous unit looked at examples of **monohybrid** crosses: one gene, two alleles, a maximum of two different types of gamete. **Dihybrid crosses** deal with two pairs of different traits simultaneously: two genes, four alleles, up to four different types of gamete. An example is shown below.

Fig. 2.5.14 shows the results of a cross between a homozygous polled black individual and a red one with horns. Fig. 2.5.15 explains allele behaviour.

In cattle, black is dominant to red.* Polled (hornless) is dominant to having horns.

* Note that other genes affect the distribution of colour in cattle, such as having white faces, or patterns of black and white.

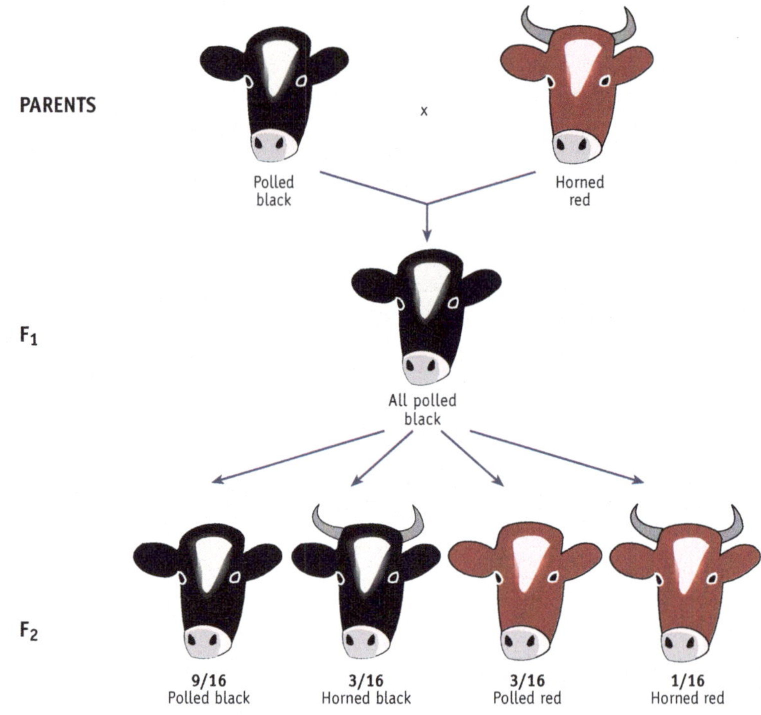

Fig. 2.5.14 Results of many matings between homozygous polled black individuals and red individuals with horns. The genes are not sex linked, so the results would be exactly the same whether the cow or the bull was black.

Hereford cattle always have a red coat and a white face. The allele for red coat is recessive, and the allele for white face is dominant. When Herefords are crossed with any other breed of cattle, their F_1 calves always have white faces.

ISBN: 9780170372855

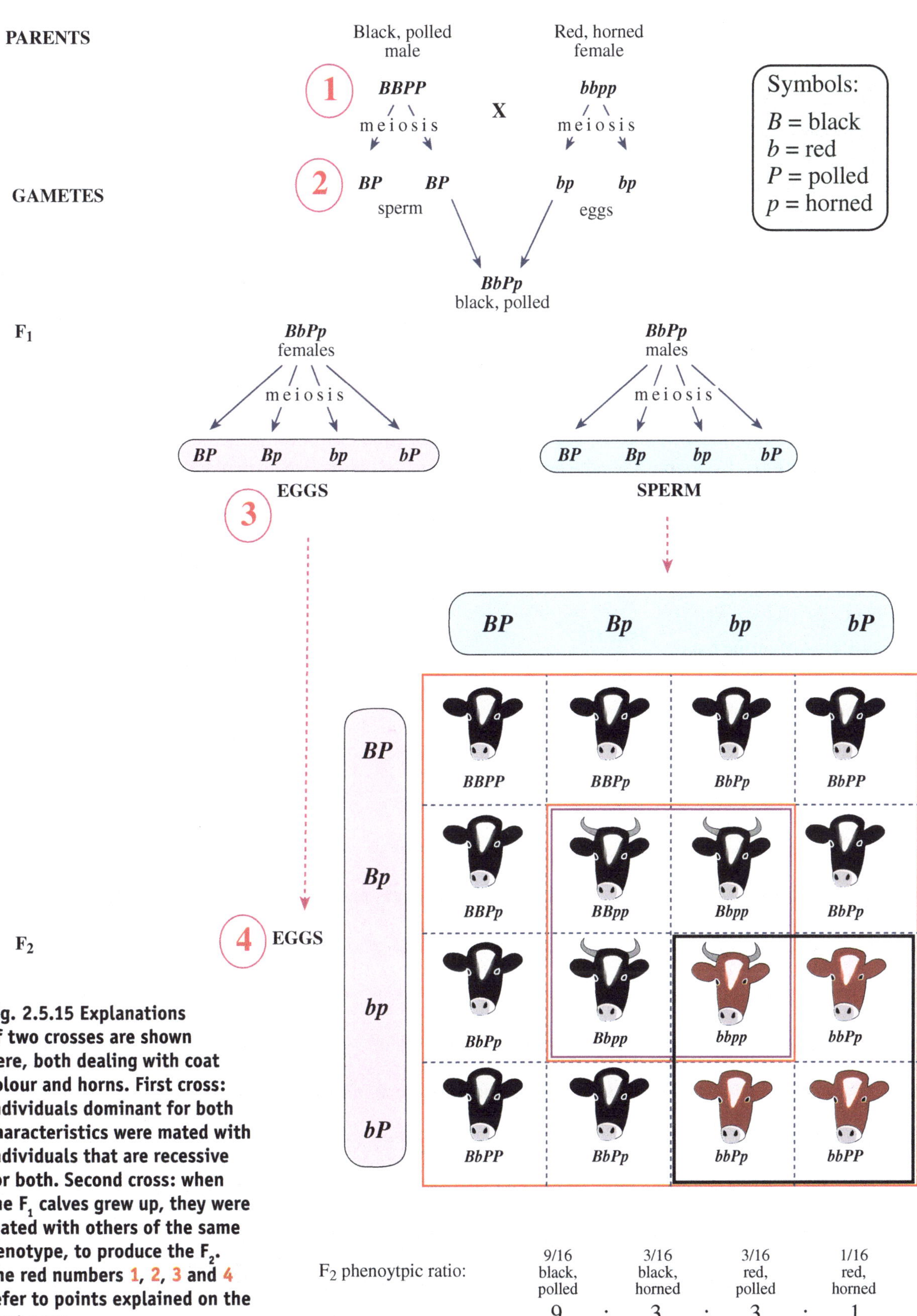

Fig. 2.5.15 Explanations of two crosses are shown here, both dealing with coat colour and horns. First cross: individuals dominant for both characteristics were mated with individuals that are recessive for both. Second cross: when the F_1 calves grew up, they were mated with others of the same genotype, to produce the F_2. The red numbers 1, 2, 3 and 4 refer to points explained on the next page.

Note: gametes can be written in any order at the side and top of a Punnett square, and the same phenotype ratios will appear. In the square above, a particular order was used as this shows a pattern of phenotypes quite clearly. 9-3-3-1 includes several 3:1 ratios.

Four points are noted in red in Fig. 2.5.15:

1 Each animal has two alleles for coat colour and two alleles for horn presence/absence.
2 Each gamete has one allele for coat colour and one allele for horn presence/absence.
3 When the F_1 cattle make gametes, half these gametes receive the *B* allele, and half of this half also receive the *P* allele. Result: a quarter of gametes carry *BP*. Altogether, four kinds of gamete are produced in equal numbers. Underlying reason: the two pairs of alleles segregate (are 'sorted') independently during meiosis.
4 Because the F_1 make four kinds of egg and four kinds of sperm (for these two genes), the Punnett square is 4 x 4. Some of the 16 compartments give the same results: the F_2 has nine possible genotypes and four different phenotypes.

A dihybrid testcross

A **testcross** is used to find whether an individual is homozygous or heterozygous. This can also be used in dihybrid situations. Example: a black cow with no horns could be homozygous dominant for both loci (*BBPP*), or it could be carrying recessive alleles (such as *BbPp*). A testcross involves mating this cow to a bull that is recessive for both traits.

An example is shown in Fig. 2.5.16. In the situation illustrated, the offspring have four different phenotypes in equal proportions, proving that the cow in question was heterozygous. But if the resulting testcross calves were to turn out all black and all polled, that would prove that the cow is homozygous dominant, and therefore suitable for 'pure line' breeding.

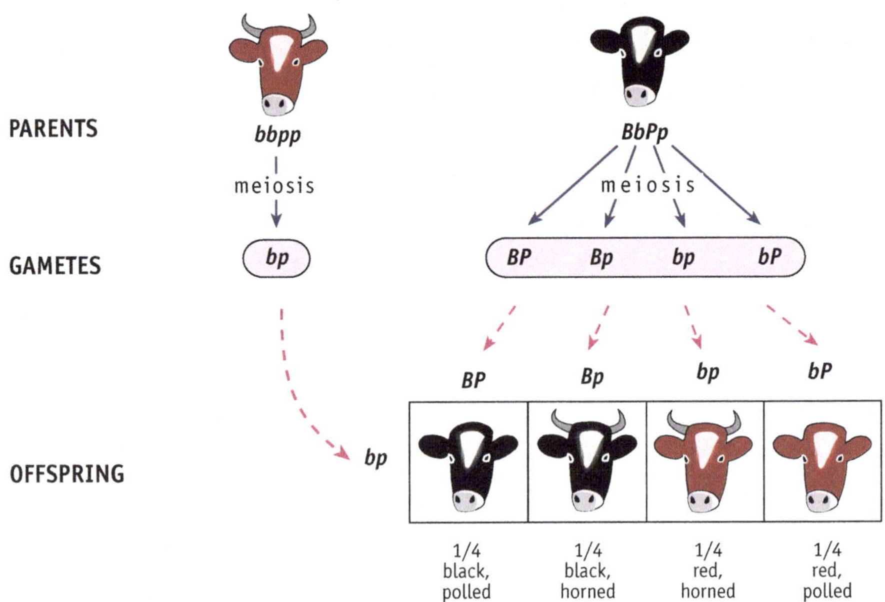

Fig. 2.5.16 A dihybrid testcross in cattle showing the likely results from a double homozygous recessive individual (*bbpp*) being mated to a double heterozygote (*BbPp*).

Angus cattle: an all-black polled breed.

Cattle with red coat and horns.

ISBN: 9780170372855

Activity A

Complete the following dihybrid cross by filling in all the blank spaces. Steps **1–6** can be used as a template for any similar situation. This cross is based on the inheritance of two pairs of different traits in pea plants, first studied by Gregor Mendel. (Mendel, 1822–84, was the first person to correctly explain inheritance patterns.)

The allele for round seeds (*R*) is dominant over the allele for wrinkled seeds (*r*).
The allele for yellow seeds (*Y*) is dominant over the allele for green seeds (*y*).

1 **Parent** genotypes and phenotypes.

Parent A: purebred plant with round yellow seeds, genotype *RRYY*

Parent B: purebred plant with wrinkled green seeds, genotype ________________

2 **Parent gametes.** Parent A: *RY*; Parent B: ______________

3 F_1 **genotypes**: all *RrYy*, and their phenotypes ________________________________

4 F_1 **gametes.** List the four kinds of gamete that will be produced by F_1 plants:

RY, *Ry*, ____________________, ____________________

5 **Punnett square.** Fill in all 16 cells.

	RY	***Ry***	***ry***	***rY***
RY	*RRYY*	*RRYy*		
Ry				
ry				
rY				

6 F_2 **phenotypes.** The expected ratios (as fractions of 16) will be:

round yellow ________________

round green _________________

wrinkled yellow ______________

wrinkled green ______________

Remember: because of meiosis, each gamete carries only one allele for each character, never both alleles.

7 Identify the one genotype whose descendants would breed true to produce round green pea seeds.

____

ISBN: 9780170372855

Activity B

Hereford cattle always have red bodies and white faces. Angus cattle always have black bodies and faces. F_1 hybrids of these two breeds always have black bodies and white faces. The phenotypes for body colour and face colour are each controlled by one gene, and these two genes are on different chromosome pairs.

1 From the evidence above, state which body colour is dominant: ____________________, and which face colour is dominant: ____________________.

2 Choose suitable letter symbols to represent each of these alleles:

red body __________, black body __________, white face __________, black face __________

3 State the F_1 genotype. ____________________

4 List the genotypes of the F_1 gametes. ______________________________

5 A cow with red body and white face is known to be heterozygous for face colour.

a State this animal's genotype. ______________________________

b List its possible gamete genotypes. ______________________________

6 This cow is mated with an F_1 bull. Complete the Punnett square below. (It's possible you will need only eight 'results' cells, not all 16.)

7 State the predicted phenotypes and phenotype ratios of calves from the kind of cross you have just used in the Punnett square.

______________________________; ______________________________;

______________________________; ______________________________;

 ISBN: 9780170372855

Dihybrid crosses with linked genes

Genes on the same chromosome are literally 'stuck together'; they are **linked**. Despite being linked, these genes often form new combinations as a result of **crossing over**, which happens in the first division of meiosis.

In *Drosophila* fruit flies, the normal ('wild type') eye colour is bright red. A mutant type has purple eyes, due to a recessive allele (symbol *pr*). Another mutation causes stunted 'vestigial' wings, due to another recessive allele (*vg*). The symbols for the wild-type dominant alleles are written vg^+ and pr^+. The loci of these genes are on the same chromosome (Fig. 2.5.18).

Because they are physically attached, linked genes do not segregate (sort) independently. As a result, we do not get the usual dihybrid ratios of 9:3:3:1 and testcross ratios of 1:1:1:1.

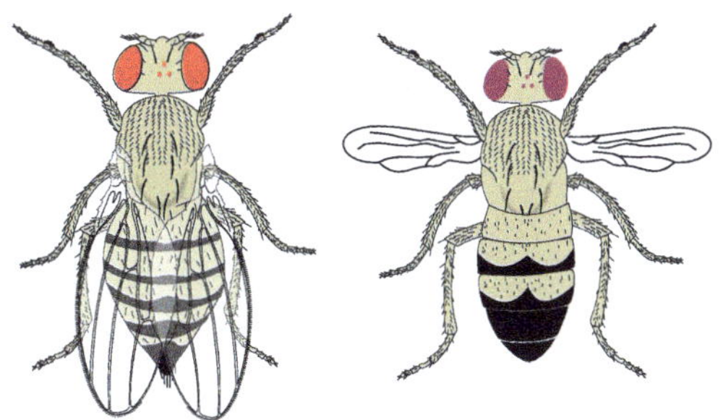

Fig. 2.5.17 *Drosophila* wild-type female (left), and vestigial-winged purple-eyed male (right).

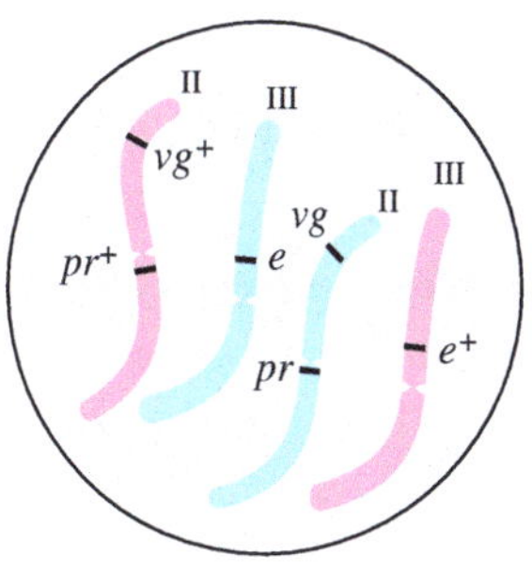

Fig. 2.5.18 *Drosophila* has four chromosomes pairs, but only two pairs are drawn here. Two genes are shown linked on chromosome II. Another gene is on chromosome III.

E

Recombinants

Fig 2.5.19 shows the result of testcrossing F_1 females. As with unlinked genes, four phenotypes are produced. The two more numerous phenotypes (67 + 62) are called **parental types** because they have the same combinations of features as their parents. The less common phenotypes (11 + 9) are **recombinants** because they show new combinations of traits that didn't feature in the parents of the F_1: 11 with vestigial wings and red eyes, 9 with normal wings and purple eyes. These 11 + 9 results can be explained by crossing over, which occurs when homologous chromosomes come together in pairs during meiosis (see Unit 2).

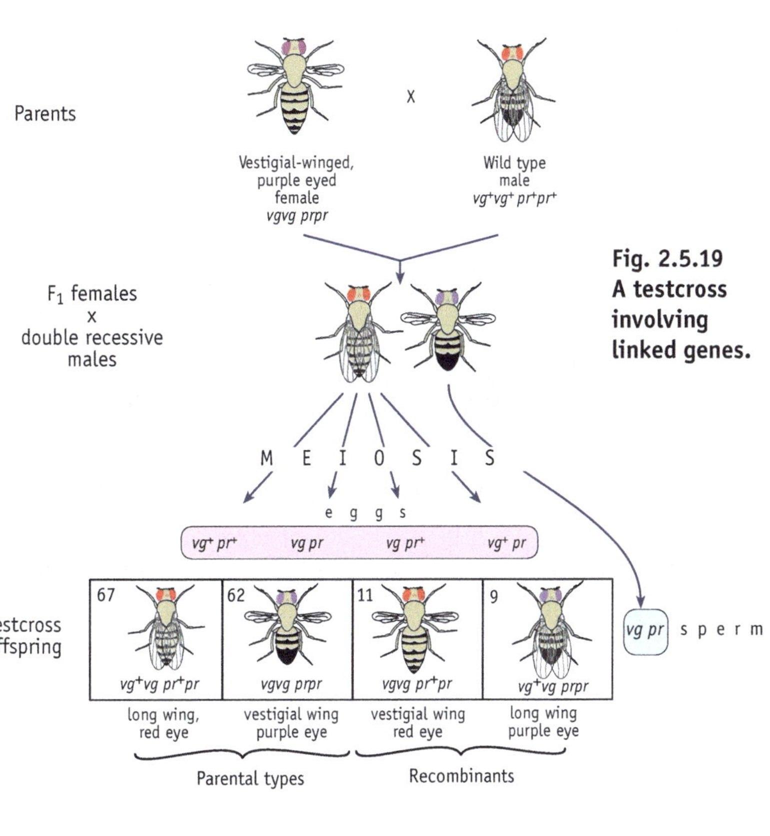

Fig. 2.5.19 A testcross involving linked genes.

ISBN: 9780170372855

If these two genes for wings and eyes were always separated by crossing over, four kinds of gametes would be produced in equal proportions, giving 1:1:1:1 ratios of phenotypes, like unlinked genes do.

Why the 11 + 9 recombinants? Crossing over does not always separate genes. Genes that are far apart on the same chromosome are more likely to have crossing over happening between them. Genes that are close together ('close loci') are less likely to be separated by crossing over and to produce recombinants.

In Fig. 2.5.19, a total of 149 flies were bred. Of these, 20 (11 + 9) were recombinants. This gives a **crossover value** (COV) of 20/149, which is 13%. A COV of 5% would indicate that two loci are close together. A COV of 40% would indicate two loci far apart.

Repeated breeding experiments like this create **chromosome maps**, showing the relative positions of genes on a chromosome.

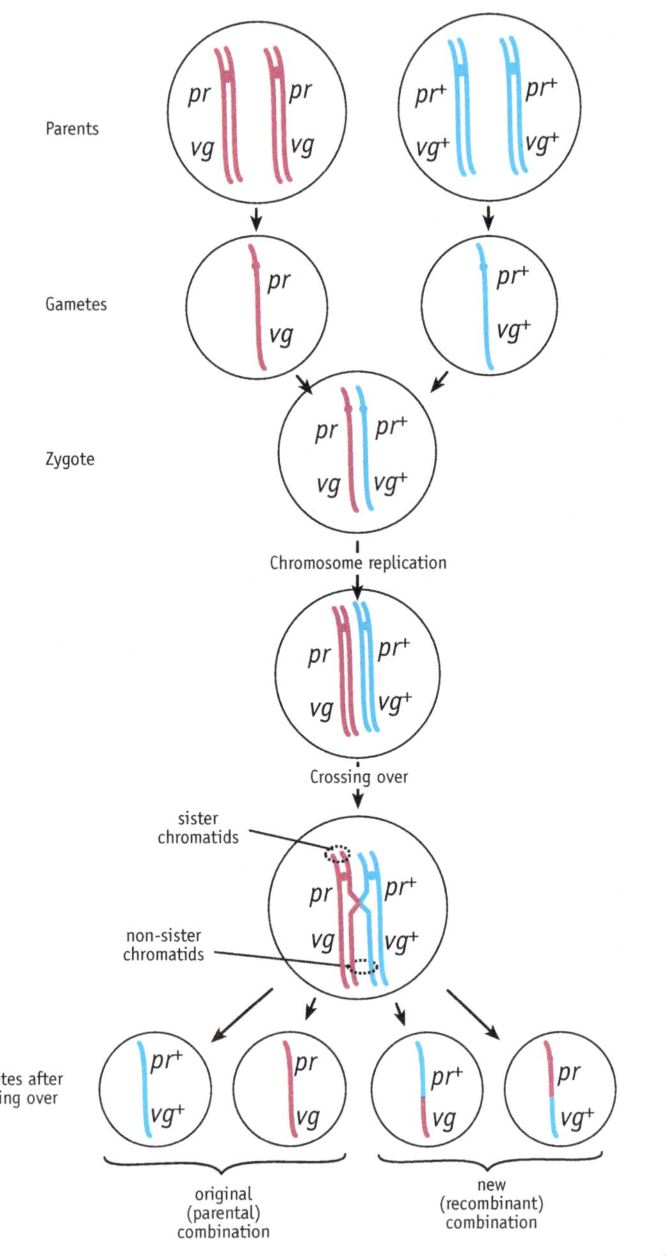

Fig. 2.5.20 Crossing over in meiosis can separate linked genes and create new gene combinations.

Activity C

1 A species of garden plant has either red or yellow flowers, and the stems can be either tall or short. Each trait is controlled by one gene, each gene having two alleles. Red is dominant to yellow, tall is dominant to short. A purebred red tall plant is pollinated by a purebred yellow short plant. The F_1 plants are all red-flowered and tall. To find out if the two genes are linked or not, some F_1 plants were testcrossed by pollinating them with short yellow plants.

a State what testcross percentages are likely if the flower and height genes are not linked.

tall red ________, short red __________, tall yellow __________, short yellow ________

 ISBN: 9780170372855

b State what testcross percentages are likely if the flower and height genes are linked and there is no crossing over.

tall red ________, short red ___________, tall yellow __________, short yellow ________

c The actual experimental results were as follows:

tall red 47%, short red 3%, tall yellow, 3%, short yellow 47%.

Explain what these results indicate.

____

____

____

____

____

2 Nail-patella syndrome (NPS) is a condition in humans where the kneecap fails to develop and the fingernails are small. It is due to a dominant allele and is not known in the homozygous condition. The NPS gene locus is on chromosome 9, which also has the locus of the gene controlling ABO blood groups. In the following pedigree, the shaded symbols represent individuals with NPS. Some genotypes for blood groups are also given.

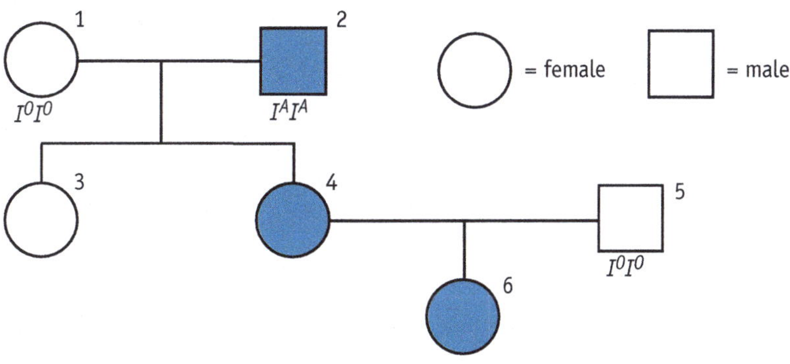

Nail-patella pedigree

a Explain what is meant by the term 'locus'.

____

____

b Identify the relationship between the gene for NPS and the gene for ABO blood group.

____

c Identify the phenotype of individual 1 for both of the traits being discussed.

____

d Predict the blood group of individual 6. Explain why you chose this blood group.

e From the information given, state one reason why NPS is *not* sex linked.

f Suggest why NPS is not known in the homozygous condition.

3 Coat colour in English setter dogs is controlled by two genes, *B* and *E*. The presence of at least one dominant allele at each locus results in black coats. Double homozygous recessive individuals have yellow coats. Individuals with the genotypes *Bbee* or *BBee* are red. Individuals with genotypes *bbEE* or *bbEe* are brown. The four different colours are the result of the gene for colour being expressed to different extents, depending on the presence of other alleles.

a State the genotype of any yellow individual.

b State the phenotype of any *BbEe* individual.

c A black female was mated to the same brown male on several occasions. They produced a total of 14 pups: 2 yellow, 3 red, 4 brown, 5 black. State the genotypes of the two parents.

d A breeder has a yellow female and wants to get a litter of only black and red pups in approximately equal numbers. State what phenotype and genotype should be chosen as a father.

e The same breeder has a brown male whose mother was yellow. This male was mated to a red female that has the same yellow mother. Predict the possible coat colours of their offspring. Justify your prediction with reasoning.

5

ISBN: 9780170372855

Unit 5 | Changes in allele frequency

Gene pools

A **gene pool** is the total of all the genes in a population. **Allele frequency** is the numerical proportion of different alleles in a population. Frequency can be measured as a percentage, or as a decimal fraction of 1. Example: a frequency of 17% is also $f = 0.17$. Calculating allele frequency is not as simple as counting the numbers of each phenotype, as Activity A (page 124) shows.

Fig. 2.5.21 Most of these birds are plain blue because they carry dominant allele *B*. A small number in this imaginary population of 50 have white patches on their wings, caused by being homozygous recessive (*bb*).

Any animal or plant population consists of an interbreeding collection of individuals within one region. The total of all their genes constitute a gene pool. Individuals die, but their genes are passed on.

ISBN: 9780170372855

Activity A

Questions **1** to **4** are based on Fig. 2.5.21.

1 The number of all-blue individuals = ______________ = ____________% of the population.

The number of white-patch individuals = ______________ = ____________% of the population.

2 The number of individuals with *BB* genotype = ________________

So their number of *B* alleles = 2x this number = ______________

The number of individuals with *Bb* genotype = ________________

So their number of *B* alleles = ________________, and *b* alleles = ______________

The number of individuals with *bb* genotype = ________________

So their number of *b* alleles = 2x this number = _______________

3 Total number of *B* alleles = ____________ out of 100, i.e. __________%, or *f* = ______________

Total number of *b* alleles = ____________ out of 100, i.e. __________%, or *f* = ______________

4 Explain why the frequency of *b* alleles calculated in question **3** is much higher than the proportion of white-wing individuals counted in question **1**.

____

____

____

Changing frequency

In many situations the frequency of different alleles in a population changes over time. Any of the following can cause frequencies to change:

- mutations (see Biology 2.5, Unit 1, and Biology 2.7, Unit 3)
- selection (both natural and artificial)
- the founder effect
- genetic bottlenecks
- immigration
- genetic drift.

The eventual effect of these processes is evolution; which means genetic change over long periods of time. This unit has information on underlying processes and mechanisms of evolution, and these are dealt with in greater detail in Biology 3.5.

 ISBN: 9780170372855

Natural selection

Selection is the main process that 'drives' changes in allele frequency. Natural selection theory is summarised in the next box.

Fact 1: All living things overproduce when breeding, and most young die before reaching breeding age.

Fact 2: There is genetic variation in every species.

Theory*: Individuals that are genetically better fitted (adapted) to their surroundings have a better than average chance of producing offspring that survive. They are 'selected for'. Their offspring inherit their genes, and these slowly become more frequent in the gene pool. Less well-adapted individuals have a lower than average chance of reproducing; they are 'selected against'. The process is known as **natural selection.**

Overall summary:
Mutations and sex both **create** genetic variation.
Natural selection slowly **removes** some variations from gene pools.

* 'Theory' is used here in the scientific meaning of 'a big idea that explains many facts'. Examples of other theories: atomic theory, plate tectonic theory.

The idea of evolution began with ancient Greeks who realised that living things change over time, but had no useful idea how this worked. In 1859 Charles Darwin published a book containing an idea for a 'working mechanism' of evolution. He called this mechanism **natural selection**, and it is summarised in the box above.

Darwin's idea began as hypothesis but it is now accepted as a confirmed theory. He had little understanding of genetics, so details of his ideas have altered — but nobody has come up with a better overall explanation. Evolution by natural selection is now the central idea of modern biology.

Natural selection is good at explaining how species evolve and adapt, how gene pools change, how new species appear. Natural selection is not about the origins of life. Science has little evidence on this.

E

Success?

In natural selection, 'success' depends mainly on reproductive success, not on survival alone. If an individual survives to old age but leaves no offspring, it has made no contribution to the gene pool.

Natural selection is not a simple matter of survival for some individuals and early death for others. One phenotype may have only a minor advantage over another, with slightly increased chances of success. Even if the average 'success' of phenotype A is 10% and for phenotype B is 7%, then genes for A will become more common in the long run. Put differently, 90% of A and 93% of B either die young or fail to breed. By human standards, natural selection is wasteful and inefficient. But it's effective.

Natural selection is sometimes known as survival of the fittest, but this can be misleading because in this context 'fit' means 'well adapted to its environment' — which is not the same as fast and strong. Many weak animals are well adapted and highly successful. Example: sparrows, snails.

Artificial selection

For thousands of years humans have altered their crops and pets and livestock to increase the numbers of those with desired features. The process is known as **artificial selection**, aka **selective breeding**. In many ways it is similar to natural selection, but differs in being goal-directed and comparatively fast.

Types of selection

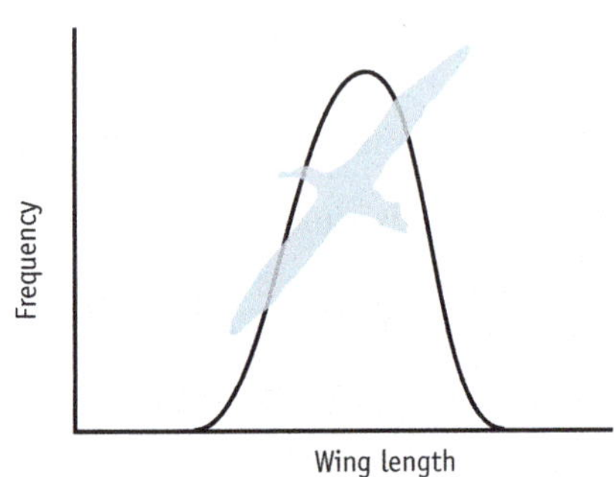

Fig. 2.5.22 In an albatross population there is a range of wing lengths.

In any population there is a range of phenotypes, usually with most individuals being close to average and with smaller numbers at each extreme. Example: wing length in albatrosses. A graph representing the situation is typically a bell curve (**normal distribution**), as in Fig. 2.5.22.

Natural selection is always acting on individuals in a population, so a phenotype graph may change over generations. Factors affecting death or breeding are often described as **selection pressure** — which can be thought of as natural selection 'forcing' a decrease or increase in one particular phenotype. Of course, this can only happen if there is lots of variation in the gene pool to begin with. The end results of selection pressure: better adaptations, as seen in Biology 2.3.

In the case of albatrosses, selection pressure may eliminate some birds with wings that are longer than average (they break) and also eliminate birds with wings shorter than average (need more energy to glide). The result: stabilising selection. Three different types of result from selection pressure are shown in Fig. 2.5.23.

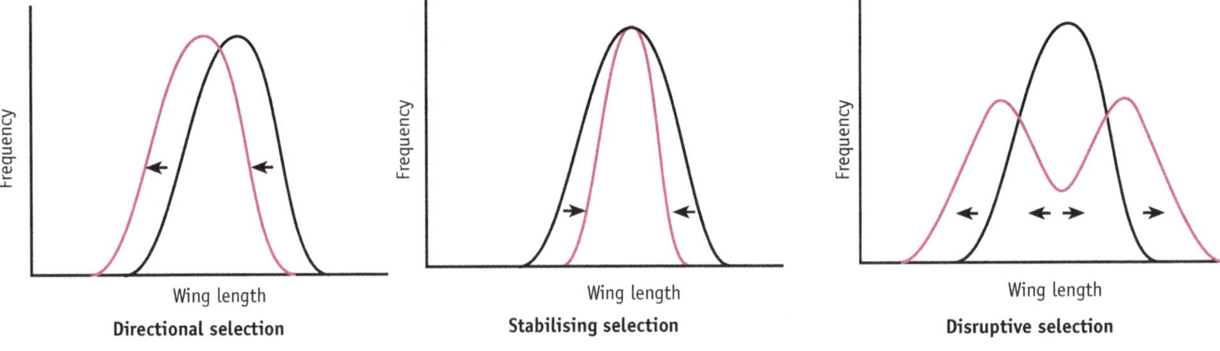

Fig. 2.5.23 When selection pressure acts against the extremes, stabilising selection causes the extremes to become less frequent. If selection pressure selects against one extreme, then directional selection will shift the distribution towards the other extreme. If selection pressure acts against individuals in the middle of the distribution, the result is disruptive selection.

The founder effect

The **founder effect** is any situation where a few individuals arrive in a region and are the founders of what eventually becomes a large population. The new population will have only those genes carried by the original few.

Example 1: Six hares were introduced into New Zealand in the 1800s, and the entire hare population is descended from these few.

Example 2: Pukeko are almost identical to an Australian species. Evidence suggests that some pukeko arrived in New Zealand about 300 years ago by flying across the Tasman Sea. It's likely this was a rare event and that the big pukeko population in New Zealand is descended from a few founder individuals.

ISBN: 9780170372855

Genetic bottlenecks

A **genetic bottleneck** is any situation in which a population crashes close to extinction, then recovers. The recovering population will have only those genes carried by the surviving few. Several New Zealand species have recovered from near extinction, and the examples named here have little genetic variety.

Example 1: Little spotted kiwi. These were heading for extinction by 1912, so five individuals were put on predator-free Kapiti Island. This saved the species but caused a genetic bottleneck that affects the more than 2000 alive now.

Example 2: Tieke (saddleback). By 1910 these were reduced to tiny populations on two offshore islands. Thanks to conservation efforts, tieke have recovered from fewer than 30 individuals to a population of more than 10,000 spread across several islands — but still carrying the genes of the original few.

Example 3: Takahe. These were once believed to be extinct, but a small population of 50 was rediscovered in the 1940s. These have now increased to almost 300 descendants.

Takahe.

Example 4: Cheetah. These African predators have little genetic variety, probably because they recovered from a population bottleneck about 10,000 years ago.

Immigration

Immigration can change the makeup of a gene pool only if both these conditions occur:

- the immigrant individuals are genetically different from the resident population, but still interbreed with them
- immigrant numbers are large. One migrant per thousand residents will have little effect on a gene pool.

New Zealand wildlife example: introduced mallard ducks often hybridise (interbreed) with the native grey duck species (parera), to the extent that the grey's gene pool has been changed and few unmodified grey individuals still exist.

Genetic drift

Genetic drift is any change in the frequency of an allele due to chance. Any chance loss of alleles is much more likely to have an effect in small populations. Imagine a population of 20 goats on a remote island; some black, some brown. In this gene pool the alleles for black and brown are equally common. In a landslide, seven goats are killed. By chance, five of the seven are brown. This accident reduces the frequency of the allele for brown colour. Once a particular allele has very low frequency, there is a high probability that further events will wipe it out completely. 'Genetic drift' is also used as a category heading that includes bottlenecks and the founder effect.

ISBN: 9780170372855

Activity B

1 Write the word(s) that match each definition or description in the table below. Select from this list: *genetic bottleneck, selection, tieke, gene pool, grey, population, founder effect, cheetah, frequency, disruptive.*

a	All the members of one species living in an area	
b	All the genes in one population	
c	A measurement of how common different alleles are	
d	A few individuals start a new population	
e	Species collapses, then recovers from a low level	
f	A mammal species affected by genetic bottleneck	
g	Endemic bird affected by genetic bottleneck	
h	Duck population genetically changed by immigration	
i	Selection pressure against average individuals	
j	Any process that affects the reproductive success of some individuals	

2 For each of the processes listed below, classify whether it tends to increase or decrease the amount of genetic variety.

a mutations ____________________

b natural selection ____________________

c the founder effect ____________________

d genetic bottlenecks ____________________

e immigration ____________________

f genetic drift ____________________

g meiosis ____________________

h fertilisation ____________________

i stabilising selection ____________________

j directional selection ____________________

3 Identify one way in which genetic bottlenecks and the founder effect are similar. Name one New Zealand example of each.

ISBN: 9780170372855

4 Identify one way in which genetic bottlenecks and the founder effect are different.

5 Predict the likely effect on selection pressure against flamingos with legs longer than average. Sketch a graph to represent the situation, and label the axes.

6 Since the 1970s, numbers of the Chatham Island black robin have recovered from the lowest possible level: one female and a few males. Suggest three reasons why the many present-day black robins are not identical to that original 'bottleneck' female.

Activity C | Hands-on

This activity imitates natural selection. Start by preparing numbers of toothpicks coloured with food dye (yellow, blue and green are suitable) in equal numbers all mixed up (perhaps 40 of each). These toothpicks represent different-coloured prey animals. Scatter them randomly across an open lawn area roughly 2 m by 2 m. At a signal, one member of your group becomes the predator and has 30 seconds to pick up as many toothpick prey as possible. Count the totals captured.

1 Record the results in this table.

Hunt 1 results	Yellow	Green	Blue
At the start	40	40	40
Captured in 30 s			
% escaping capture			

5

ISBN: 9780170372855

2 If many toothpicks remain uncaught, arrange another 30-second hunt for the 'survivors'.

Hunt 2 results	Yellow	Green	Blue
At the start	?	?	?
Captured in 30 s			
% escaping capture			

3 Repeat with a fresh lot of toothpicks on different grass; different colour or length to the first trial.

Hunt 3 results	Yellow	Green	Blue
At the start	40	40	40
Captured in 30 s			
% escaping capture			

4 Form conclusions on what these results tell you about natural selection.

5 Suggest two or three ways in which this experiment is not realistic.

5

ISBN: 9780170372855

Biology 2.5 Genetic variation and change

AS 91157 Demonstrate understanding of genetic variation and change
Externally assessed, 4 credits

Achievement	Achievement with Merit	Achievement with Excellence
Demonstrate understanding of genetic variation and change.	Demonstrate in-depth understanding of genetic variation and change.	Demonstrate comprehensive understanding of genetic variation and change.

Achievement
'Demonstrate understanding ...' involves defining, using annotated diagrams or models to describe, and describing characteristics of, or providing an account of, genetic variation and change.

Achievement with Merit
'Demonstrate in-depth understanding ...' involves providing reasons as to how or why genetic variation and change occurs.

Achievement with Excellence
'Demonstrate comprehensive understanding' involves linking biological ideas about genetic variation and change. The discussion of ideas may involve justifying, relating, evaluating, comparing and contrasting, or analysing.

Notes

1 'Genetic variation and change' involves the following concepts:
- sources of variation within a gene pool
- factors that cause changes to the allele frequency in a gene pool.

2 'Biological ideas' and processes relating to sources of variation within a gene pool are selected from:
- mutation as a source of new alleles
- independent assortment, segregation and crossing over during meiosis
- monohybrid inheritance to show the effect of co-dominance, incomplete dominance, lethal alleles, and multiple alleles
- dihybrid inheritance with complete dominance
- the effect of crossing over and linked genes on dihybrid inheritance.

3 'Biological ideas' and processes relating to factors affecting allele frequencies in a gene pool are selected from:
- natural selection
- migration
- genetic drift.

4 Examination: Candidates may be required to interpret diagrams and new information, and write responses of one or more paragraphs. Some questions may be resource-based. Candidates are required to understand the difference between gametic and somatic mutations, as sources of new alleles. Candidates may be required to draw and/or interpret Punnett squares and calculate the expected proportions of genotype and phenotype (expressed as a ratio, fraction, percentage or decimal). Genetic drift is considered to include founder effect and bottlenecks.

Revision 3 | Genetic variation and change

Write a list of key points in each box.

Chromosomes, genes, alleles

Mutation

Meiosis

Monohybrid

Dihybrid

Gene frequency, alleles

5

ISBN: 9780170372855

Revision 4 | Genetic variation and change crossword

Across

1 The position on a chromosome occupied by a gene (5)
6 Having different alleles for a given gene locus (12)
8 Describes an allele that is only expressed in homozygotes (9)
9 Having three of a particular chromosome (7)
12 Having two of the same allele (10)
14 The joining together of gametes (13)
19 Cell formed by fusion of two gametes (6)
20 Describes variation in which there are distinct kinds (13)
21 An organism's observable characteristics (9)
22 A variant of a particular gene (6)
23 Nuclear division that produces variety (7)
24 Process in which gene pool is greatly reduced as a result of temporary population crash (10)

Down

2 Point at which homologous chromosomes interlock after crossing over (7)
3 Kind of selection that resists change (11)
4 Mechanism by which new alleles are produced (8)
5 To have any importance in evolution, variation must be at least partly … (7)
7 Process in which genes on same chromosome pair are recombined (8, 4)
10 Chromosome not concerned with determination of sex (8)
11 All the genes in a population (4, 4)
12 Chromosomes that pair in meiosis are said to be this (10)
13 The genetic makeup of an organism (8)
15 Method of distinguishing between homozygotes and heterozygotes (9)
16 Kind of selection that acts against one end of a range of variation (11)
17 Having more than two sets of chromosomes (9)
18 Kind of selection that acts against the middle of a range of variation (10)
20 Syndrome resulting from extra copy of chromosome 21 in humans (4)

5

ISBN: 9780170372855

Exam-type question 1

Sex linkage

Haemophilia is an inherited disease of the blood in which it fails to clot, so the sufferer may bleed to death after even a minor injury. Haemophilia figured prominently in the history of European royalty in the 19th and 20th centuries. Queen Victoria had nine children with Prince Albert, who did not have the disease, and none of her ancestors had it. Two of her four daughters (Princess Alice and Princess Beatrice) passed the disease to various royal houses across the continent, including the royal families of Spain, Germany and Russia. Victoria's son Prince Leopold suffered from the disease. For this reason, haemophilia was once popularly called 'the royal disease'.

The X-linked condition manifests itself almost entirely in males because it is recessive in females, although the gene for the disorder may be inherited from either mother or father. The alleles are X for normal clotting and X^h for the haemophilia allele.

Discuss haemophilia in the royal family including the following.

- Explain what a 'sex-linked disorder' means, giving another example.
- Explain, giving reasons, how Queen Victoria may have acquired the allele to pass to her children.
- Complete the Punnett square to show the possible children she *could* have had with Prince Albert.
- Why the disease is more common in males than in females.
- Why not all of her five sons got the disease.

Use this scaffold framework as the basis for an answer to be done on your own paper.

- Define 'sex linked' and explain what 'disorder' means.
- Name an example of a sex-linked trait, **other than** haemophilia.
- Suggest how the trait could have appeared in Queen Victoria. Refer to the stimulus material: none of her ancestors had haemophilia, nor did Prince Albert.
- Explain why you think the trait had arose in this way, and why not through inheritance.
- Fill in the Punnett square, labelling Queen Victoria's genotype and also Prince Albert's.
- Write the predicted ratios for possible offspring.
- Explain why this trait (haemophilia) appears more often in males than in females. Refer to which chromosome it is carried on.
- Explain what is meant by recessive and dominant alleles.
- Explain why although Queen Victoria had five sons, some of them did not get haemophilia.
- Use the Punnett square to help justify her genotype, and also to show how she could have produced sons who carried a chromosome without the allele.

ISBN: 9780170372855

Exam-type question 2

The inheritance of coat colour in horses is quite complex. The palomino colour is derived from crossing a Chestnut (CC) with a White($C^{CH} C^{CH}$).

The image shows a palomino mare with her foal.

A breeder wishes to produce a herd of only palomino horses.

Discuss the breeding of palomino horses including:

- a Punnett square illustrating how to produce only palomino foals,
- how the herd would be maintained as only palomino,
- a reason why this is an example of incomplete dominance and not co-dominance,
- a description of an example of co-dominance.

Use this page to create a key points plan or a mind map or scaffold that could form the basis for a longer answer to be done on your own paper.

ISBN: 9780170372855

Exam-type question 3

Manx cats

Tailless cats were known by the early 19th century as cats from the Isle of Man, where they remain a substantial but declining percentage of the local cat population. The taillessness arose as a natural mutation on the island though folklore persists that tailless domestic cats were brought there by sea. They are descended from mainland stock of obscure origin and the situation is therefore described as a founder effect. Like all house cats, including nearby British and Irish populations, they are ultimately descended from the African wildcat (*F. silvestris lybica*) and not from native European wildcats (*F. s. silvestris*), of which the island has long been devoid.

The recessive allele *m* codes for a normal tail and spine development.

The dominant gene *M* codes for a short/stumpy tail and normal spine.

The pure Manx cat is not possible as in the homozygous condition it is fatal, with the spine failing to develop in the foetus and none is born alive.

Discuss the Manx cat genetics, addressing the following in your answer.

- Why the 'Manx' condition can be attributed to the founder effect.
- Why the Manx allele became common on the Isle of Man.
- What other genetic implications this has for the Manx cats on the island.
- Complete the Punnett square to show the result of a cross between two Manx cats.
- Describe the expected genotypic and phenotypic ratio of the offspring.

Use this scaffold framework as the basis for an answer to be done on your own paper.

- Define 'founder effect'.
- Explain why this definition fits the situation in Manx cats. (Refer to the stimulus material.)
- Explain how Manx cats became common on this island. Refer to population size, and also to factors that influence gene pools.
- Explain why in this case it matters whether the allele is dominant or recessive.
- Suggest how the gene pool makeup could be altered if some offspring die.
- Explain which ones are likely to die, and what effect this will have on surviving offspring ratios.
- Since the Manx cat population is small, explain what may tend to happen over time to the allele frequencies of the total genome (in addition to the Manx-tail allele).
- Explain whether the situation would tend to give the Manx cat population greater genetic diversity or less genetic diversity.
- Would all alleles continue in the gene pool? Explain.
- Predict what might happen if closely related Manx cats breed. What is this situation called?
- Complete the Punnett square. Label all parts correctly.
- Clearly indicate the ratio of expected offspring.

ISBN: 9780170372855

Exam-type question 4

The inheritance of cat coat colour is very complex. Some colours are due to one gene that has many alleles. The alleles vary in determining the amount of pigment distributed down the hair shaft.

C = full brown coat all over and is dominant to every other variation
c^b = Burmese colour pattern: brown coat and with green eyes
c^s = Siamese colour pattern: cream with brown 'points' (ears, paws and tail)
c^a = albino with blue eyes
c = albino with red eyes

If the Burmese and Siamese alleles appear together, they are known as co-dominant and the resulting animal is a Tonkinese: dark brown in a Siamese pattern, but with intermediate turquoise eyes. Both of these characters are dominant over albino. Blue-eye albino is dominant to red-eye albino.

A breeder wanted to buy two cats that when crossed would produce four different colours in their kittens. Discuss how the breeder would go about doing this. Base your discussion on a completed version of the Punnett square below. Include:

- the genotype and phenotype of the parents
- the genotype and phenotype ratio of the expected offspring
- the chance of producing a Tonkinese in the litter
- any problems that may be encountered in the health of the cats, due to their rarity.

Female gametes

Male gametes		

Use this page to create a key points plan or a mind map or scaffold that could form the basis for a longer answer to be done on your own paper.

ISBN: 9780170372855

Exam-type question 5

Cavalier King Charles spaniels come in two different colours: black and tan (*E*) or ruby (*e*).

There is a second gene locus that determines whether the coat is solid (*S*) or broken with white (*s*).

This gives a combination of four different colour varieties:
Ruby – solid red-coloured coat
Black and tan – distributed as black and tan over the whole body as in a Doberman pinscher
Blenheim – rich red broken up by large white areas
Tricolour – the black and tan is broken up by white (as in the Blenheim).

A breeder has a Blenheim female and wants a male that would allow her to produce all four phenotypes.

Discuss how the dog breeder would go about ensuring that all colour phenotypes were possible in the puppies.

- Describe the genotype and phenotype of the male dog.
- Use a Punnett square to work out the possible genotypes of the offspring.
- Describe the expected phenotype ratio of the offspring.
- Explain why she may not produce all the different types of coat colour possible from this cross.

Use this scaffold framework as the basis for an answer to be done on your own paper.

- Write the genotypes for all four colour varieties described in the stimulus material.
- Write the breeding male's genotype, and also describe what his coat colour (phenotype) must look like. Note: Pups of all phenotypes are desired, so the breeding male's genotype must include all alleles, since the female is double recessive for colour, and heterozygous for pattern.
- Explain how any mating will not necessarily produce the theoretical predicted ratios.
- Complete the Punnett square. Check that it shows pups with all four possible phenotypes. (Note: Place female on left side, as she is homozygous for colour, so has fewer alleles combinations.)
- Write the genotype and phenotype for each offspring.
- State what Punnett squares are used for.
- Refer to the chance of particular genotypes egg and sperm meeting.
- State whether or not each litter of pups is influenced by ratios in the previous litter.
- State whether or not the number of pups makes a difference to the resulting colour ratios.

ISBN: 9780170372855

Exam-type question 6

In *Drosophila* fruit flies, grey bodies (*G*) are dominant to ebony (*g*), and normal wings (*N*) are dominant to vestigial (*n*).

Discuss crossing over as it affects linked genes. In your answer you should:

- describe what is meant by linked genes
- explain the process of crossing over of linked genes; you may illustrate your answer with a diagram
- explain the effect of crossing over on genetic variation; use *Drosophila* fruit flies as an example.

Use this page to create a key points plan or a mind map or scaffold that could form the basis for a longer answer to be done on your own paper.

ISBN: 9780170372855

Exam-type question 7

In 2011 an albino native land snail *Powelliphanta* was found in Kahurangi National Park. It had a pigment-less foot and head.

This is remarkable because *Powelliphanta* snails usually have a dark brown coloured body.

Consider the above information in your answer to the following.

- Explain the role of mutation in evolution.
- Explain the process of natural selection.
- Define genetic diversity and explain its importance.

Use this scaffold framework as the basis for an answer to be done on your own paper.

- Describe the effect that mutations have on the number of alleles for a trait, and hence on variation in a gene pool.
- State whether mutations always cause a change in phenotype. Explain the significance of this.
- Name some different types of mutation. Since any mutation may cause a change in protein, specify how each type of mutation could harm (or be useful to) the individuals that receive them.
- State some specific examples of mutations; link your examples to colour in snails.
- Discuss natural selection with regard to these snail colours.
- White snails are more visible. What implications does this have?
- Brown snails are less visible. What implications does this have?
- Discuss what could happen (over time) to the numbers of snails of each colour.
- Suggest factors that may reduce or increase the number of both colours. This could include predation, and also the effect of melanin in most organisms in sunny conditions.
- Define genetic diversity.
- Explain the evolutionary significance of population size. (Genetic drift, etc.)
- Explain the implications to a whole species if its populations are small and have little genetic diversity.
- Referring to these snails, suggest the long-term effects if diseases arise, or the climate changes.

 ISBN: 9780170372855

Exam-type question 8

Strawberry plants reproduce by both sexual and asexual methods. They produce short runners that remain attached to the plant, and grow roots from nodes that grow into the ground as soon as they touch it and also start producing leaves. These offspring are identical to the parent plant. However, they can also produce flowers that are pollinated by bees and the fruit that results from fertilisation has seeds on the outside. These seeds may then be eaten and distributed by birds or humans.

Discuss the reproductive strategy of strawberry plants by:

- explaining how sexual reproduction can increase variation
- outlining the advantages of having two types of reproduction.

Use this page to create a key points plan or a mind map or scaffold that could form the basis for a longer answer to be done on your own paper.

ISBN: 9780170372855

Exam-type question 9

The black robin or Chatham Island robin (*Petroica traversi*) is an endangered species from the Chatham Islands off the east coast of New Zealand. Although it can fly its flight capacity is somewhat reduced. Its evolution in the absence of mammalian predators made it vulnerable to introduced species such as cats and rats, and it became extinct on the main island of the Chatham group before 1871, being restricted to Little Mangere Island thereafter. The species is still endangered, but now numbers around 250 individuals in populations on Mangere Island and South East Island. Ongoing restoration of habitat and eradication of introduced predators is being undertaken so that the population of this and other endangered Chatham endemics can be spread to several populations.

'Small populations are much more prone to changes in the gene pool because of genetic drift and migration.'

Discuss this statement relating your answer to the black robin and include the following.

- Define these terms: gene pool, gene flow and migration.
- Explain how the black robin's gene pool could change over time.
- Explain how a small population differs from a large one when considering changes in a gene pool.
- Any implications to the black robin of having a small population in terms of the species future.

Use this scaffold framework as the basis for an answer to be done on your own paper.

- Define gene pool.
- Define gene flow.
- Define migration.
- Read the stimulus material carefully, then suggest how the particular environmental conditions of New Zealand today could be impacting on the black robin population.
- Is the robin gene pool small or large? Explain whether the present size of their population today is likely to lead to genetic drift.
- Define genetic drift.
- Describe how allele frequency in black robins is likely to change due to chance.
- Give an example of a chance event that could impact this population.
- Explain why small and large populations differ in the likelihood of their allele frequencies changing.
- List a number of factors that can affect populations, and explain the effect of each. Refer to the definitions asked for in the question.
- Explain how today's black robin population size and genetic diversity are likely to influence whether the species is likely to continue. Refer to future (named) changes to the environment.
- Finally, suggest what humans could do to help the black robin population recover.

5

ISBN: 9780170372855

Exam-type question 10

The takahe is a rare New Zealand flightless bird. It was thought extinct until discovered in the Murchison Mountains of the South Island. Originally its habitat extended over both main islands but since the recent introduction of predators such as stoats, ferrets and rats, the population plummeted and is now thought to be only about 250 birds. A few birds have been moved to predator-free islands such as Tiritiri Matangi and Maud Island in the Marlborough Sounds.

- Discuss the terms population bottleneck, genetic drift and founder effect in relation to the takahe.
- Include the long-term implications of moving small numbers of birds to islands and any steps that will need to be taken to ensure the long-term survival of the species.

Use this page to create a key points plan or a mind map or scaffold that could form the basis for a longer answer to be done on your own paper.

ISBN: 9780170372855

Ecology

NCEA Achievement Standard 91158

Investigate a pattern in an ecological community, with supervision

2.6

Internally assessed, 4 credits.

Unit 1 | Communities

Natural 'wild' systems can be found almost anywhere, even in a little patch of plants. The following words are used when describing any such system:

- **habitat**: any place with plants and animals living in it; their 'address'. A habitat could be 20 cm diameter, or 20 km.
- **niche**: the way of life for any particular species; its 'occupation'.
- **community**: all the living things in a particular habitat. A community can be as simple as a few species living in a cave, or as complex as a coral reef.
- **food web**: the food-energy links between living things in a community.
- **ecosystem**: all the living organisms in a community — plants, animals, decomposers — together with their energy links, all chemicals, and the recycling of nutrient chemicals.
- **ecology**: scientific study of any of the above.

A range of at least eight habitats can be seen in this photo. Each habitat has its own distinct community of plants and animals and micro-organisms.

ISBN: 9780170372855

Activity A

First, identify eight different habitats in the photo on page 154. Next, draw a neat indicator line from each of these areas to its matching description above the photo.

Biotic and abiotic

A community may have patterns such as:

- **zonation**: changes in plant and animal types at a boundary between two habitats
- **stratification**: vertical layering in plant communities in tall forest
- **succession**: any process of community change that happens over many years. Succession, also known as regeneration, usually follows a major disturbance such as fire, erosion, landslip, overgrazing.

Fig. 2.6.1 An outline of a forest interior and edge. The arrow represents zonation; the brackets show stratification.

Pioneer plants are still establishing themselves on the bare rocks of Rangitoto Island, centuries after it stopped erupting. Once a few pioneers take hold, their fallen leaves begin to make soil and their shade begins to create conditions more suitable for other plants. This in time leads to the next stage in the succession process. The eventual result will probably be a mature forest containing completely different plants to the original pioneers.

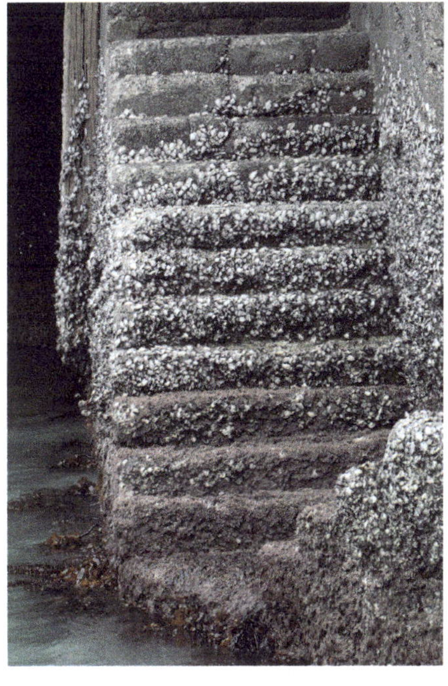

Intertidal zonation. Seaweeds (steps 1–4), oysters (steps 4–11) and mussels (step 11) form visible zones here.

The factors affecting living things in ecosystems can be grouped in two broad categories.

Biotic factors. Includes predators, food (prey), competitors, parasites, social interactions.

Abiotic factors. Includes all physical and chemical factors such as sunlight, wind, water, rainfall, temperature, chemicals in soil and water. (*A-biotic* means 'not living'.)

Two major abiotic factors for creatures living on intertidal rocks:
(a) the force of waves, which could dislodge them;
(b) being exposed to dry air for several hours at each low tide.

6

ISBN: 9780170372855

Activity B

1 Review some basic concepts in ecology by writing the word(s) that match each definition or description in the table below. Select from this list: *stratification, soil moisture, habitat, abiotic, tapeworm, competitors, organism, community, food web, zonation, succession, ecosystem.*

a	Any place with plants and animals living in it	
b	All organisms living in a particular habitat	
c	All linked components in a habitat: living, physical, chemical	
d	A diagram showing who eats what	
e	A general word for any living thing	
f	Non-living factors that affect living things	
g	Changes in species mix at the edge of a community	
h	One example of an abiotic factor	
i	Vertical layering in plant types in a forest	
j	One example of a parasite	
k	Animals that use or need the same food supply	
l	Community changes that take place over time	

2 The following factors are known to influence seal populations. Classify each of the factors as either biotic or abiotic.

a Ocean temperatures at the time. ______

b Presence or absence of orca (killer whales) in the area. ______

c Abundance of fish in the area. ______

d Abundance of krill that these fish feed on. ______

e Day length and the amount of sunlight. ______

f Rival males on the breeding beach. ______

g Whether the seal population is high or low at the time. ______

h Storms and ocean waves. ______

i Polluting chemicals in the water. ______

 ISBN: 9780170372855

Choosing a community

Achievement Standard 91158 requires you to investigate a pattern in an ecological community. Details are given at the end of Unit 2.

Whichever community is being investigated, you will need to analyse and interpret information about it. 'Patterns' may include succession, zonation, stratification, seasonal changes — and you must consider some biotic and/or abiotic factors. Any of these factors could affect the patterns in the community. You could include results from a 'natural environment' study obtained as part of Biology 2.1. Your teacher will give advice on how to present your report and how much time you have.

Your teacher may decide which community is to be studied, or may leave this decision to you. Many different communities could be suitable, such as:

- a natural forest
- rocky shore intertidal zone
- any relatively undisturbed habitat in the school grounds, or at home
- a lawn area containing a mixture of several kinds of plants
- marshland or a lake shore
- stream or pond.

Easy access is one reason to choose a habitat within the school grounds. Lawns are less suitable for study as they often have only one or two plant species and the animals may be hard to find.

A small patch of unused ground, less than a metre across but containing several plant and insect species, all interacting as parts of an ecosystem. Example: the fungus depends on the dead branch for food, and eventually recycles nutrients back into the soil.

Activity C

Complete the table below by writing one or two major advantages or disadvantages in each box. Factors to consider when deciding which community may be more suitable for investigation:

- how easy or difficult it will be to get to the place
- how much time you will be able to spend there
- how difficult or easy it will be to identify and count different species.

Community	Advantages	Disadvantages
1 Natural forest		

ISBN: 9780170372855

Community	Advantages	Disadvantages
2 Rocky shore intertidal zone		
3 Undisturbed weedy area		
4 Lawn		
5 Pond		
6 Another area of your choice		

6

ISBN: 9780170372855

Unit 2 | Sampling methods

Questions about natural communities fall into two broad categories.

- **Types and numbers.** What species of plants and animals are living here? What are the total numbers of each? What are the numbers of males and females and of different age groups?
- **Distribution patterns.** Is each species spread randomly, or is it patchy, or in zones? Are the patterns linked to soil type, light, vertical stratification, etc? Is distribution of one species linked to another's? In what ways is distribution reflected in adaptations? Is the population showing signs of changing over time? Why?

It's impossible to count and measure every individual plant and animal, no matter how small the community being investigated. However, it is possible to get a good overall picture by **sampling**, which means collecting data in a few selected areas. Total numbers in the sample areas can then be scaled to give population estimates for the whole community.

Sampling: where to count

Three common ways of choosing sample areas are:

1. **Straight line transects.** These are useful when investigating horizontal zonation. Examples: across an intertidal zone, or at a boundary between wetland and grassland.
2. **Regular grid.** Suitable for finding patterns of distribution across an area.
3. **Random.** Good for estimating the total population of a particular plant or animal, but less suitable for finding area patterns.

Sample areas (quadrats) can be square or else circular. Size could be as big as 20 m x 20 m (e.g. in a forest with large trees), down to 20 cm x 20 cm for small animals and plants.

Whichever sampling method is used, the first aim is to do no harm. Leave each habitat as you found it.

1 *Straight line transect*

2 *Regular grid*

3 *Random*

6

Activity A

1 For each of the following five situations, state which sampling method you would choose: grid, transect, random, or any of these.

a The distribution of mature kauri trees across an entire forest. ____________________

b The zoning of barnacles between high and low tides on the rocky shore. ____________________

c Seasonal changes in kiwi numbers in a forest. ____________________

ISBN: 9780170372855

d East-west changes in plant type across tussock grassland. ______________________

e The pattern of distributions of cockles across a tidal mudbank. ______________________

2 As part of an ecology project, the numbers of cockles was counted in a muddy tidal zone. A wire-frame quadrat 20 cm x 20 cm was used, and five separate counts were made. The numbers of cockles counted in the five quadrats were: 6, 5, 3, 6, 4.

a Calculate the average number of cockles per square metre.

____

____

b Calculate the average number of cockles per hectare in that habitat. (1 hectare = 100 m x 100 m)

____

____

c When you did these calculations in **2**, what assumption(s) did you make about cockle distribution?

____

____

Sampling: what to count

It is best to make counts on-site, using a prepared blank results table. When uncertain of the name of any plant or animal, take a photograph for later identification. Direct counts are easy in the case of plants and non-moving animals like oysters, but counting is less easy for animals that are shy or nocturnal or otherwise hard to see. In these situations, any of the following indirect methods (A–H) can help with estimates and counts.

A The amount of plant damage. Suitable for animals like caterpillars.
B Calls. Suitable for noisy animals.
C Footprints.
D Droppings/faeces. Suitable for possums, etc.
E Pitfall traps (Fig. 2.6.2).
F Tullgren funnel (Fig. 2.6.3, page 162).
G For insects in bushes, spread a plastic sheet on the ground, shake the branches.
H Aerial drone photography, and later counting the images on screen.
I Direct counting, where the animals/plants are easy to see and not mobile.

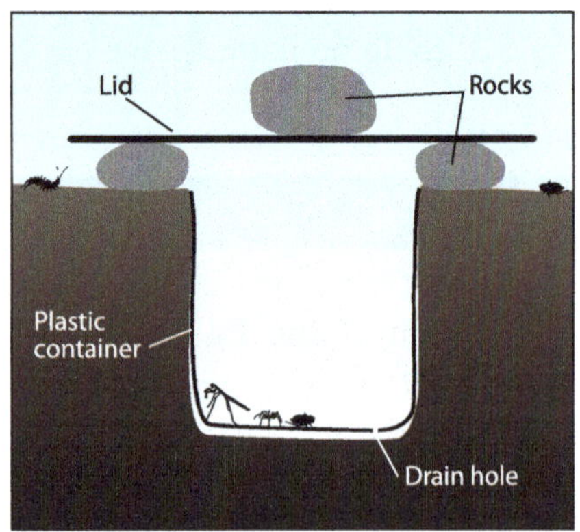

Fig. 2.6.2 Pitfall traps are suitable for sampling animals that live at ground level.

6

ISBN: 9780170372855

Activity B

For each of the following situations **1–8**, state which of the counting methods A–I (page 160) you would use. For **1–4**, justify your choice with a reason.

1 The amount of possum damage in a forest. ________

Reason: ______________________________

____

____

2 Changes in the numbers of kiwi in 10 km² forest area over several years. ________

Reason: ______________________________

____

____

3 The total numbers of elephants across 10,000 km² of grassland. ________

Reason: ______________________________

____

____

4 The numbers of barnacles on a rocky shore. ________

Reason: ______________________________

____

____

5 The numbers of deer in a forest. ________

6 The numbers of possums in a forest. ________

7 The numbers of lizards in a habitat of rocks and tussock grass. ________

8 The numbers of caterpillars per tree in an apple orchard. ________

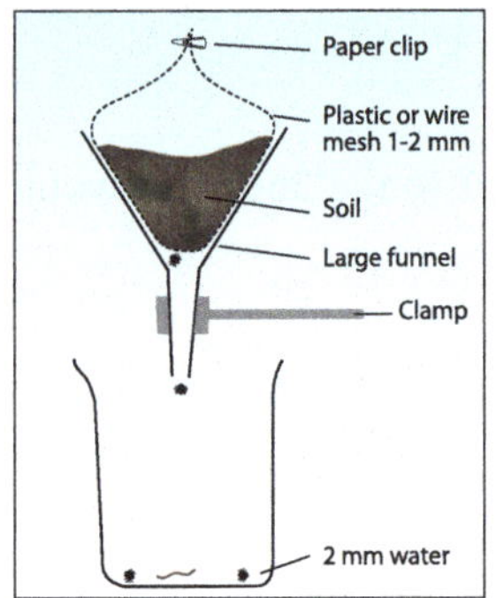

Fig. 2.6.3 A Tullgren funnel is suitable for finding small animals in soil. As the soil dries out over several hours, small animals move downwards and fall into the water. Alternatively, search for these animals by spreading the soil and leaves onto a white plastic tray, then sort through it using a small brush.

Centipedes are one of the top predators in leaf litter and soil communities. (Be careful: they can give a nasty bite.)

Activity C

The framed area on page 163 represents an imaginary forest with 100 trees, each colour being a different species. (Alternatively, the symbols could be taken to represent four different animal species on a rocky shoreline.) Your task is to take random samples by randomly dropping a glass coverslip onto the 'forest', each time recording the number of trees in the quadrat. Decide what to do in cases where a tree is partly outside a quadrat, or when the whole quadrat falls partly outside the area. Instead of a square coverslip, you could sample using a plastic ring about 2 cm diameter.

1 Record your results in this table.

Tree species:	Red	Blue	Green	Black
Total from 5 samples				
Total from 10 samples				
Total from 15 samples				
Actual numbers in the whole area				

2 Evaluate how well results from this random sampling method reflect the total actual numbers in the whole area.

3 Describe distribution patterns you can see for each of the four kinds of tree: red, blue, green, black. (The random sample method probably did not detect these patterns.)

6

ISBN: 9780170372855

4 Suggest what sampling method would be more likely to detect these patterns.

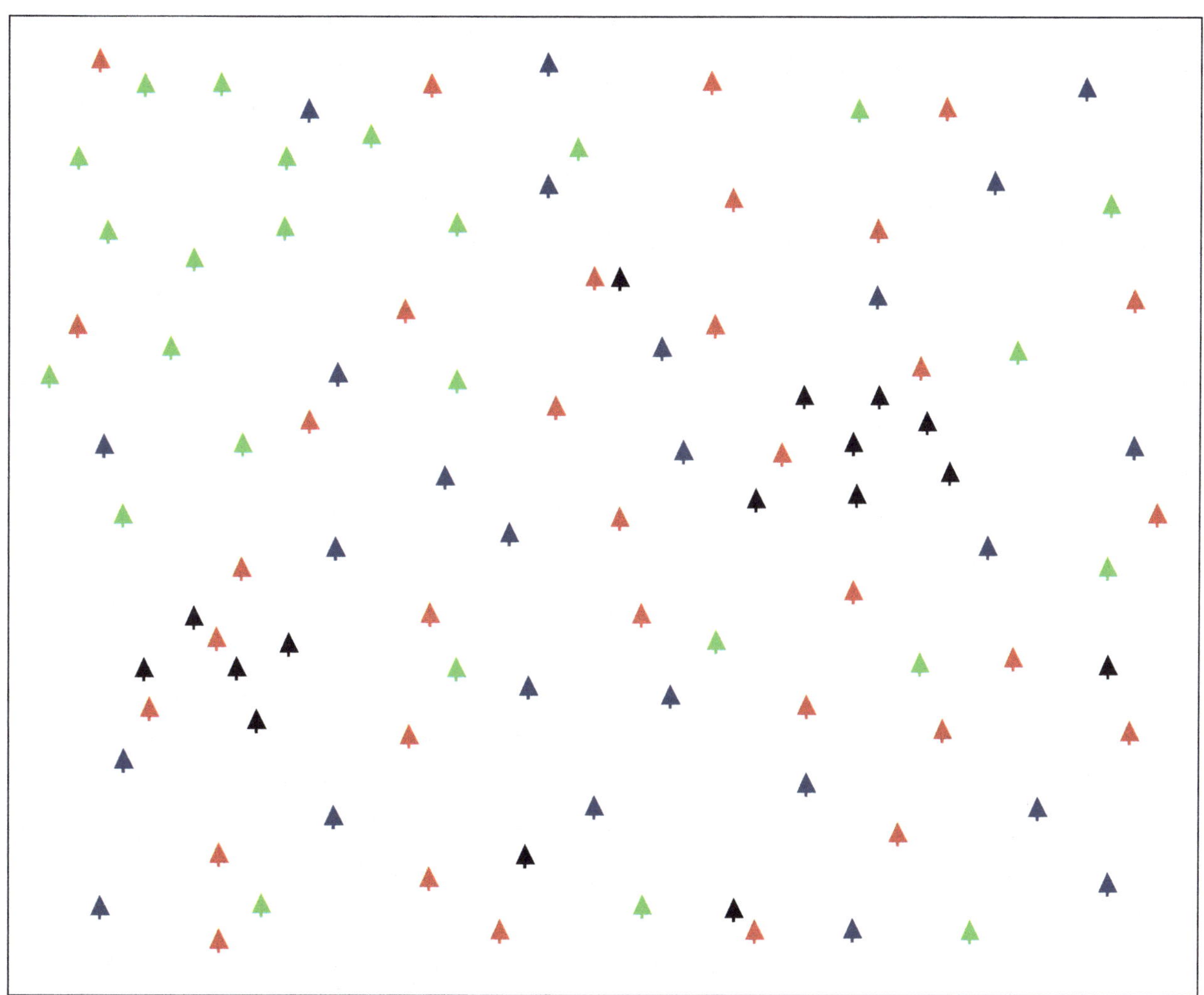

ISBN: 9780170372855

Biology 2.6 Ecology

AS 91158 Investigate a pattern in an ecological community, with supervision
Internally assessed, 4 credits

Achievement	Achievement with Merit	Achievement with Excellence
Investigate a pattern in an ecological community, with supervision.	Investigate in-depth a pattern in an ecological community, with supervision.	Comprehensively investigate a pattern in an ecological community, with supervision.

Achievement with Merit
'Investigate ...' involves describing observations or findings, and using those findings to identify the pattern (or absence of a pattern) in an ecological community, relating this pattern to an environmental factor, and describing how the environmental factor might affect chosen species within the community.

Achievement with Merit
'Investigate in-depth ...' involves providing a reason for how or why the biology of one of the chosen species relates to the pattern (or absence of a pattern). The biology involves structural, behavioural or physiological adaptations of the organism which are related to the environmental factor and to an interrelationship with an organism of another species (e.g. competition, predation, or mutualism).

Achievement with Excellence
'Investigate comprehensively ...' involves using an environmental factor and the biology of interrelated organisms of different species to explain the pattern (or absence of a pattern). The explanation may involve elaborating, applying, justifying, relating, evaluating, comparing and contrasting, and analysing.

Notes

1 Investigation involves analysing and interpreting information about the ecosystem. The information may come from direct observations, collection of field data, tables, graphs, resource sheets, photographs, videos, websites, and/or reference texts.
2 A community pattern may include: succession, zonation, stratification, or another distribution pattern in response to an environmental factor.
3 Environmental factors likely to affect patterns in a community include abiotic and/or biotic factors.
4 'Biology' of the organisms refers to any adaptations of organisms that relate to the pattern being investigated and may include interrelationships such as competition, predation, or mutualism.
5 Assessment against this standard may be based on a stand-alone or an individual investigation that can contribute findings to a larger group or class investigation. In a group or class investigation, individual findings may be discussed and individual students may interpret their own findings in the light of other students' investigations and findings. Findings from outside the group or class such as published information or historical findings relevant to the investigation may also be used.
6 It is intended that this investigation be carried out with supervision. This means that the teacher provides guidelines for the investigation such as the context for the investigation, instructions that specify the requirements for a comprehensive investigation, and broad conditions such as the availability of equipment and/or resource material. Students then develop and complete the investigation from the initial guidelines given by the teacher. Supervision may involve discussion between teachers and individual students in order to clarify the students' ideas and may also involve teachers managing the process of sharing findings.

ISBN: 9780170372855

Gene expression

NCEA Achievement Standard 91159

Demonstrate understanding of gene expression

2.7

Internally assessed, 3 credits.

Unit 1 | Protein chemistry

Protein significance

Humans are about 65% water, with more than half the other 35% being protein. There are many kinds of protein molecule with many different functions, including thousands of different enzymes. Almost all enzymes are proteins, and each enzyme controls a specific body reaction. What controls enzymes? Answer: DNA. Before looking at how DNA does this, we need to know more about proteins.

The wide range of protein functions includes:

- **enzymes**; many kinds, each of which catalyses (speeds up) a specific process
- **structural** proteins; examples: keratin (hair, fingernails) and collagen (ligaments, tendons, skin, cartilage)
- all **antibodies**; the immune system depends on these working well
- some **hormones**; example: insulin
- **oxygen carriers**; example: haemoglobin
- **movement proteins**; muscle proteins myosin and actin together make more than half of total body protein.

Hair consists of keratin. Collagen is a major part of skin, and also of cartilage in the ear.

Amino acids

All protein molecules are **polymers**: chains of many amino acids units joined together, the amino acid molecules being **monomers**. (Greek *poly*: many; *mono*: one; *meros*: part.) In proteins there are 20 kinds of amino acid — but you don't need to memorise their names.

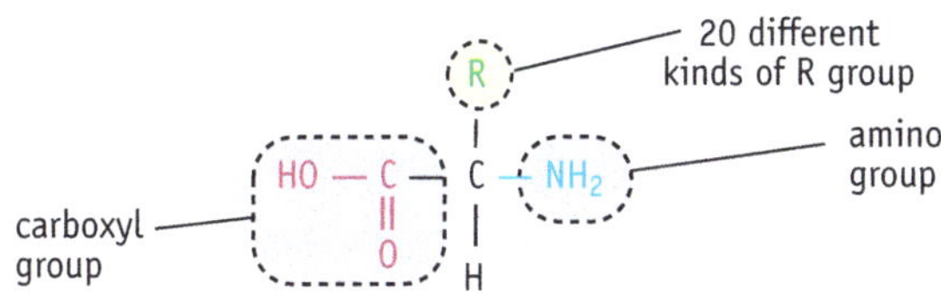

Fig. 2.7.1 A general structure of amino acids. The grey-shaded parts are the same for all 20 kinds.

ISBN: 9780170372855

Amino acids are medium-sized organic molecules, each having between 10 and about 30 atoms. All have these three parts, as shown in Fig. 2.7.1:

- one **amino group** (–NH_2)
- one **carboxyl acid group** (–COOH)
- one **R group**, a short 'side branch' of C and H atoms. Different R groups give each amino acid its individuality.

When two amino acids link up, one molecule of water is removed in the process, the resulting link being known as a **peptide bond**. Any medium-length chain of amino acids is described as a **polypeptide**. Most proteins are commonly made of several polypeptides joined in complex ways. The process of joining two smaller molecules (and producing water), as shown in Fig. 2.7.2, is known as a **condensation reaction**. The reverse process is **hydrolysis**.

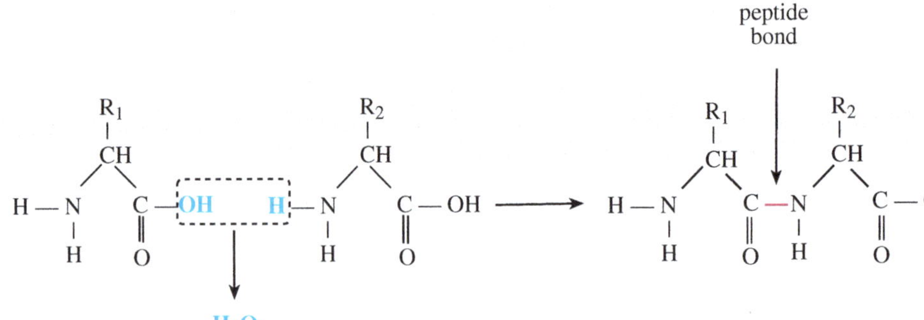

Fig. 2.7.2 A dipeptide formed by linking two amino acids. Water is a by-product.

Getting it right

Assembling a protein is not simply a matter of joining the correct number of amino acids; they must be joined in the right sequence. If even one of the 146 amino acids that make up a beta-polypeptide of a haemoglobin molecule is incorrectly placed, then the haemoglobin may not work. There is an astronomical number of ways of joining 146 amino acids. Example: there are 20 x 20 x 20 different ways of making a chain of just three amino acids — that's 8000 ways, 7999 of which are 'wrong'. With four amino acids, the number of arrangements goes up to 20^4 = 160,000. Conclusion: DNA control over protein production must be highly accurate.

Activity A

1 Give the chemical names of five different kinds of protein, and also state the function or location of each. What do these protein names have in common?

7

ISBN: 9780170372855

2 Write the word(s) that match each definition or description in the table below. Select from this list: *monomer, condensation, polymer, hydrolysis, peptide bond, haemoglobin, keratin, polypeptide, myosin, tripeptide.*

a	Molecule consisting of many units joined together	
b	Small number of amino acids joined together	
c	Individual units that make up a polymer	
d	Bond linking two adjacent amino acids	
e	Three linked amino acids	
f	Protein in hair and fingernails	
g	Reactions that join molecules and produce water	
h	Protein specialised for carrying oxygen	
i	Reactions that split larger molecules, using water	
j	One kind of muscle protein	

3 How many kinds of amino acid are there? Name one example.

____

4 Calculate the number of possible sequences for five amino acids ____________________

and for seven amino acids ____________________.

Protein shapes

Protein molecules are of two general shapes (Fig. 2.7.3):

- **fibrous proteins** with rope-like molecules. Examples: keratin and collagen
- **globular proteins**, in which peptide chains are folded into an irregular ball shape. Examples: enzymes and most other proteins.

The sequence of amino acids in a protein chain or polypeptide chain is its **primary structure**, and governs its properties. Proteins also have a **secondary structure**, in which the polypeptide chain forms either a zigzag or a helix. In addition, globular proteins have a **tertiary structure**, in which the peptide chain is folded into a ball shape.

Their complex folded shapes make most kinds of protein

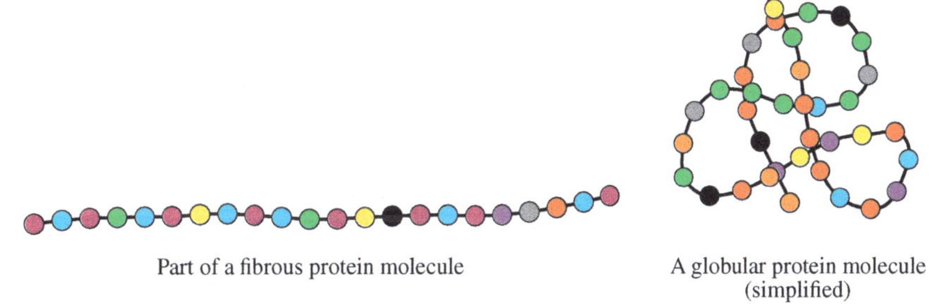

Fig. 2.7.3 The two general protein shapes. Different types of amino acid are represented here by different colours.

ISBN: 9780170372855

molecules very fragile compared with other organic substances. This is because they are held in shape by weak forces (hydrogen bonds, see below). This is why proteins are easily **denatured**, which means their tertiary structure is easily destroyed by high temperatures or changes in pH. Denaturing is permanent. You can't uncook an egg.

E

How globular proteins keep in shape

In an enzyme, only a few of the amino acids — those at the active site — actually take part in the reaction. These may be a long way apart in the chain, but are brought close together by the folding of the tertiary structure and held in position by weak chemical bonds (Fig. 2.7.4).

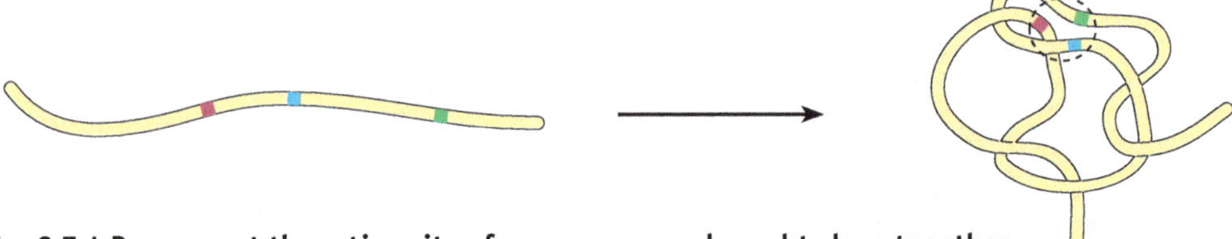

Fig. 2.7.4 R groups at the active site of an enzyme are brought close together by folding of the polypeptide chain. Active group shown by a dotted line.

The pattern of folding is specific for each kind of protein. Globular proteins are kept in shape by a similar mechanism to that which causes phospholipids to form layers in plasma membranes. The mechanism depends on water, and water molecules are polar, as Fig. 2.7.5 shows.

Some amino acids have polar R groups, which make them attracted to water by weak **hydrogen bonds**. The polypeptide chain is flexible enough to bend into any shape, but tend to adopt the shape in which the polar groups face towards the water, with the non-polar groups buried inside (Fig. 2.7.6). A similar mechanism anchors globular proteins in cell membranes. Some globular proteins also have a **quaternary** structure in which two or more polypeptide chains are held loosely together. Example: haemoglobin consists of two alpha and two beta polypeptide chains.

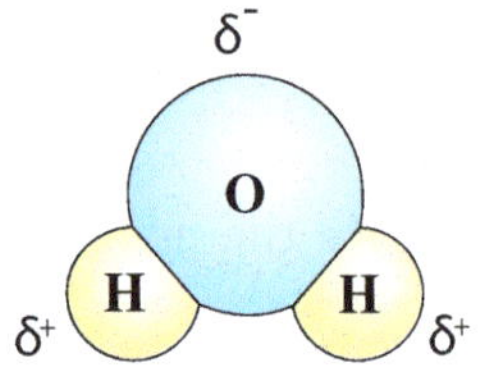

Fig. 2.7.5 Every water molecule has slight positive charges at the H atoms, and a slight negative charge on the O atom. Molecules like this are described as 'polar'.

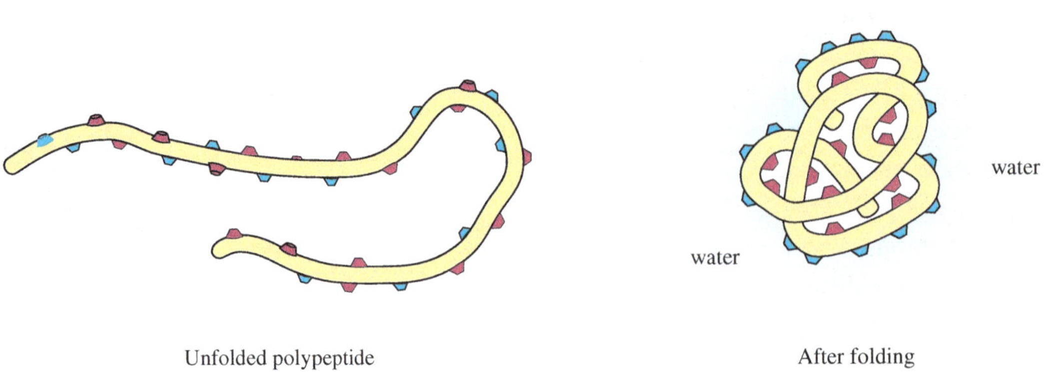

Fig. 2.7.6 In this globular protein, the peptide chain is folded so that the water-attracting polar R groups (blue) face outwards, and the non-polar groups (red) face inwards, where there is no water.

ISBN: 9780170372855

Activity B

1 Describe what is meant by protein 'primary structure'.

2 Describe what is meant by protein 'secondary structure'.

3 Identify the main feature that gives any globular protein its unique properties.

4 Name two examples of fibrous proteins, and one example of a globular protein.

5 Define what is meant by 'denatured'.

6 Explain why many kinds of protein are so easily denatured.

7 Explain what is meant by an enzyme's 'active site'; also briefly describe what keeps any enzyme molecule in shape.

ISBN: 9780170372855

Unit 2 | Assembling proteins

DNA revision

DNA controls how proteins are assembled, which makes this a good point to revise basic facts about DNA. If unsure of the facts, first check Biology 2.4, Unit 4.

Activity A

Supply missing words **1** to **13**, then complete diagram **14**.

In eukaryotes, DNA is part of the **1** ____________________, which are inside the **2** ____________________. DNA is a long, double **3** ____________________-shaped molecule, a polymer made up of thousands of monomers known as **4** ____________________. Each of these has three parts: a phosphate group, a **5** ____________________ ____________________, and a base. The two strands of DNA are held together by weak **6** ____________________ bonds between bases. There are four kinds of base in DNA: thymine always pairs with **7** ____________________, and **8** ____________________ always pairs with **9** ____________________. Before **10** ____________________ division, DNA replicates (= **11** ____________________) itself. The system known as **12** ____________________ base-pairing ensures that both new DNA double helix copies have exactly the same sequence of **13** __________ bases as the original.

14 Use the base-pairing rules to complete the lower part of the DNA strand drawn below. Write A, T, C or G in each box. The bases are grouped in threes for reasons that will be explained later.

G	C	A		T	C	G		A	C	C		T	A	G		C	A	T

7

 ISBN: 9780170372855

Protein synthesis stage 1: transcription

DNA is more than a giant molecule with an ability to copy itself.

DNA's function is to store coded information on how to build proteins.

This information is in the form of particular sequences of bases that carry instructions for particular amino acids.

Protein synthesis happens in two stages: transcription and translation.

Stage 1: **transcription** (copying). Inside the nucleus, **DNA** is copied to make **mRNA**.
Stage 2: **translation**. mRNA moves out of the nucleus to the **ribosomes**, where mRNA regulates the sequence in which amino acids are assembled to make a polypeptide. **tRNA** is a carrier molecule that brings individual amino acids to the assembly site. The whole 'assembly' process is known as **protein synthesis**.

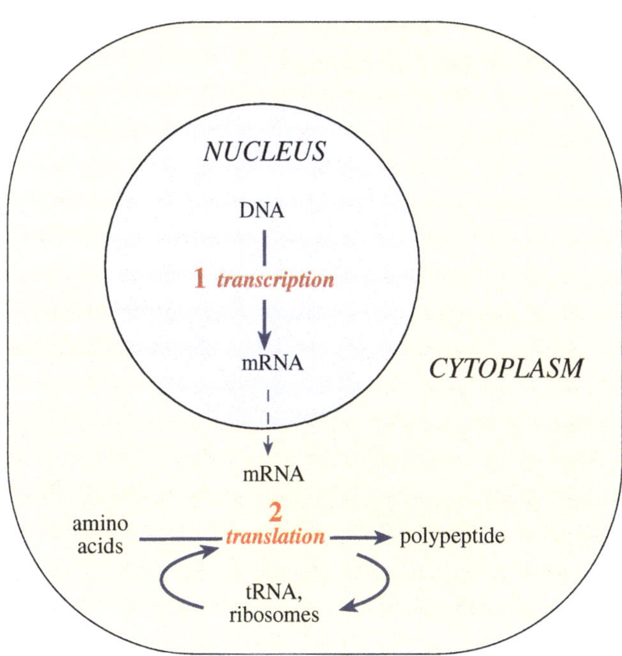

Fig. 2.7.7 The two main stages of protein synthesis.

The flow of information can be summed up as:

In some ways, this flow of information is comparable to the stages that take place when a document is printed. In both cases the 'red' information is temporary. When it has done its job, mRNA is dismantled. Likewise the printer's memory is wiped. In both cases the 'blue' information remains unchanged and stays where it was at the start.

A gene is a long sequence of bases that codes for one kind of polypeptide. (Also see E box on page 187.)

Because it depends on base pairing, the transcription stage is much like what happens in DNA replication before cell division. But there are several features that are unique to the process of transcription.

- Only a section of DNA is copied — **one gene** at a time.
- Only **one strand** of DNA is copied. This is called the coding strand of the DNA and is built on the complementary strand — known as the template strand.
- mRNA molecules are **temporary** — dismantled after use.

The 'm' in **mRNA** stands for 'messenger' because it carries a message from the nucleus to the ribosomes. **tRNA** is a related substance with its own specialised function of transferring amino acids: the 't' stands for 'transfer'.

ISBN: 9780170372855

There are similarities between DNA and both kinds of RNA, but also the following chemical differences:

- RNA has the base **uracil** in place of thymine, so the base pairing is A-U and C-G
- RNA has **ribose sugar** in its nucleotides — hence the full name ribonucleic acid.
- RNA is **single-stranded**.

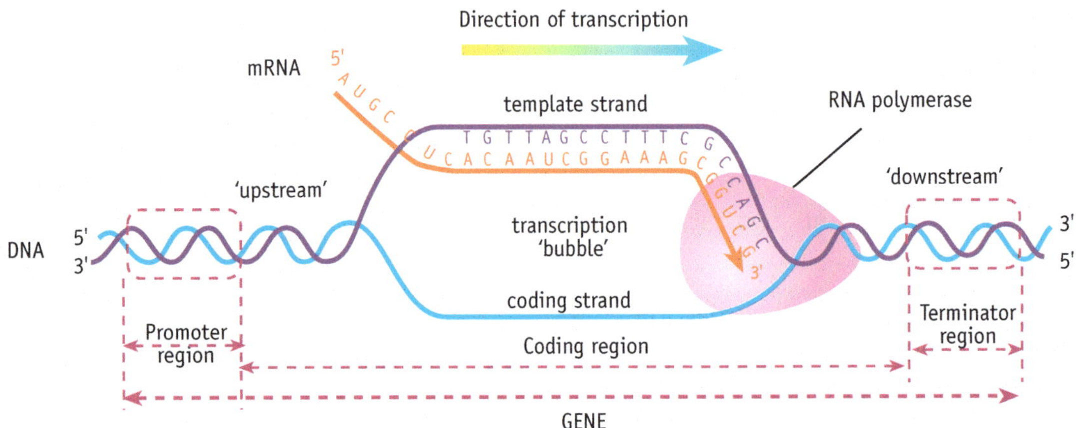

Fig. 2.7.8 Shows DNA being transcribed into mRNA. The pink shape is RNA polymerase, an enzyme responsible for joining the newly arrived nucleotides that make up the growing RNA strand. A single gene is represented here, with three regions: promoter, coding, and terminator. In reality all three are much longer than shown.

E

Transcription details

During transcription, mRNA grows only at one end. Because of details in the nucleotide sugar component, this is known as the 3′ end of mRNA. Fig. 2.7.8 shows 3′ as the leading end and 5′ as the lagging end. The complementary sides of a DNA double strand are of course different, so how does a cell know which DNA strand to use? And how does it know to copy the correct section of DNA? A DNA strand may be millions of bases long; a gene only a few thousand bases.

The template strand is identified by a particular base sequence, which is 'recognised' by RNA polymerase. Another base sequence, called the promoter, is situated just before the beginning of each gene, and is also 'recognised' by mRNA, indicating that this gene is to be copied. Copying stops when it reaches a 'terminator' sequence at the end of the DNA gene.

Protein synthesis stage 2: translation

7

Many years of research have shown that each group of three bases 'matches' one particular amino acid. In DNA language, each group of three bases is known as a **triplet**, and in mRNA language it is a **codon**. The **genetic code** is this system of three bases corresponding to each amino acid. Examples, in mRNA 'language':

- the codon **GCA** always codes for the amino acid **alanine** (ala)
- the codon **UGU** always codes for the amino acid **cystine** (cys)
- the codon **UAA** always codes for **stop** — 'end of message'.

ISBN: 9780170372855

This genetic code has been investigated in many living things from bacteria to jellyfish, from cabbages to camels — and appears to be the same in all of them. (A very few minor differences have been found, but the code is otherwise universal.) This code system must have evolved very early in the history of life on Earth for it to be universal.

Although the genetic code is universal, the 'messages' vary. You have the same genetic code as almost every living thing on the planet, but that doesn't make you genetically the same as wombats and palm trees. Their genomes differ.

However, among humans the genetic messages are very similar. It's been found that all humans ever tested have genomes that are more than 99.5% identical.

Activity B

Complete the following sentences **1** to **5**, each in seven to 15 words.

1 As regards DNA length, a gene is ______________________________

____

____

2 Transcription is a process in protein synthesis in which ______________________________

____

____

3 Translation is a process in protein synthesis in which ______________________________

____

____

4 The role of RNA polymerase is to ______________________________

____

____

5 A codon consists of ______________________________

____

____

7

6 (Refer to Fig. 2.7.10) The base sequences of the three 'stop' codons are

____

7 Complete this table to summarise differences between DNA and mRNA.

Feature	DNA	mRNA
Single strand, or double?		
Name of sugar component?		
Comparative length?		
Nucleotide bases? (letters only)		
Summary of function? (in fewer than 15 words)		

Reading a codon table

7

There are 64 different codons. You don't need to memorise these, but you need to be able to use a table to find which amino acids are specified by which codons.

Look at the left side of the table and read across the panel of the first base, then look at the top and read down the column of the second base. Finally, use the right of the table for the third base. This will give you the letters identifying a particular amino acid. Example: AGA leads to 'arg' (arginine).

Note that three codons are marked STOP, meaning 'end of sequence'. AUG indicates START and also doubles for the amino acid 'met' (methionine), so every polypeptide begins with this amino acid — although it is usually deleted immediately.

Examples of coding systems invented by humans: alphabets, digital codes, Morse code, musical notation.

ISBN: 9780170372855

First base of codon (5' end)	Second base of codon: U	C	A	G	Third base of codon (3' end)
U	phe	ser	tyr	cys	U
U	phe	ser	tyr	cys	C
U	leu	ser	STOP	STOP	A
U	leu	ser	STOP	trp	G
C	leu	pro	his	arg	U
C	leu	pro	his	arg	C
C	leu	pro	gln	arg	A
C	leu	pro	gln	arg	G
A	ile	thr	asn	ser	U
A	ile	thr	asn	ser	C
A	ile	thr	lys	arg	A
A	met + START	thr	lys	arg	G
G	val	ala	asp	gly	U
G	val	ala	asp	gly	C
G	val	ala	glu	gly	A
G	val	ala	glu	gly	G

Key: Blue: Amino acids with R groups that are attracted to water
Pink: Amino acids with R groups that are not attracted to water

Fig. 2.7.9 The genetic code table, expressed in RNA codons. This could also be written in DNA coding strand triplets. Note that four codons are 'punctuation marks'.

E

Why three?

RNA has four kinds of base: A, U, C, G. If each codon had only two bases, there would be 16 different possible combinations. Since there are 20 different amino acids, 16 codons would be inadequate to differentiate between them. With four different bases and three bases per codon, there are 64 different possible combinations (4 x 4 x 4). Four bases per codon would give 256 possible combinations — far too many. Three bases is the smallest possible codon size for 20 different amino acids.

64 codons is more than enough, which is why most amino acids have at least two different codons and one has six. This situation with 'spare' codons is known as **redundant**, and the code system is sometimes described as **degenerate**. An equivalent would be a person using multiple phone numbers for the same purpose.

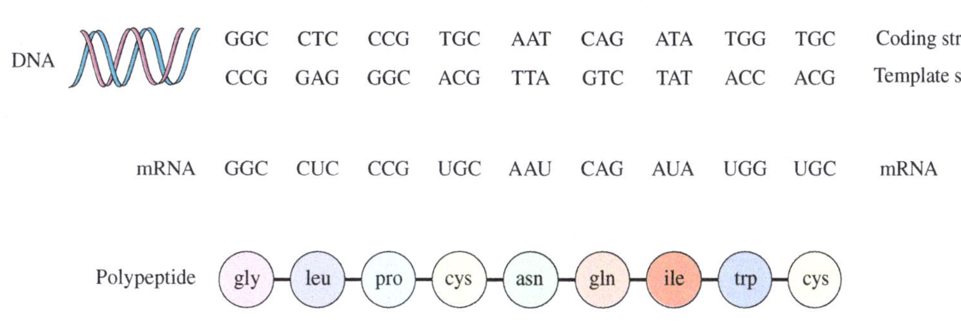

Fig. 2.7.10 Each amino acid in the hormone oxytocin is represented by a sequence of three bases in DNA and mRNA.

7

Transfer RNA, ribosomes, translation

After being produced in the nucleus, each mRNA molecule is transported into the cytoplasm for the translation stage. Making a polypeptide involves **tRNA** (transfer RNA) and **ribosomes**. Transfer RNA is a small single-stranded molecule, folded back on itself as Fig. 2.7.11 shows. It acts as a carrier and has two important features:

- One end can combine with a particular amino acid.
- Part of the molecule has another three unpaired bases called an **anticodon**. This is complementary to a particular codon on RNA.

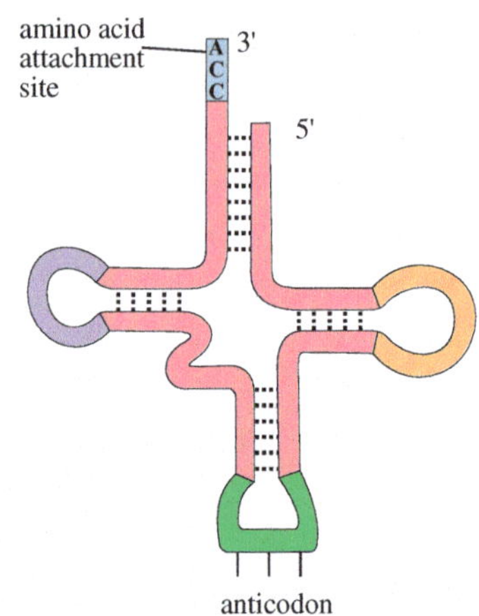

Fig. 2.7.11 A tRNA molecule showing its two important sites.

Once an RNA molecule has combined with its own particular amino acid, it combines with the corresponding codon on RNA — but can only do this with the help of a ribosome.

Ribosomes are tiny granules, particularly abundant in cells that produce large amounts of protein. A ribosome's behaviour during protein synthesis is summarised in these eight steps as Fig. 2.7.12 shows. At the start, tRNA molecules are already present, each carrying its own particular amino acid.

1 The ribosome combines with the first codon on the mRNA molecule.
2 The mRNA's codon links with the tRNA's anticodon.
3 The ribosome moves along to the next codon on the RNA.
4 A second amino acid is brought into place by its tRNA.
5 A peptide bond is formed between the two amino acids.
6 Each previously attached tRNA breaks free to combine with another amino acid.
7 The ribosome moves along another codon, the process being repeated until it reaches a stop codon.
8 Finally, the ribosome breaks free and can begin to translate another mRNA of the same gene or another gene.

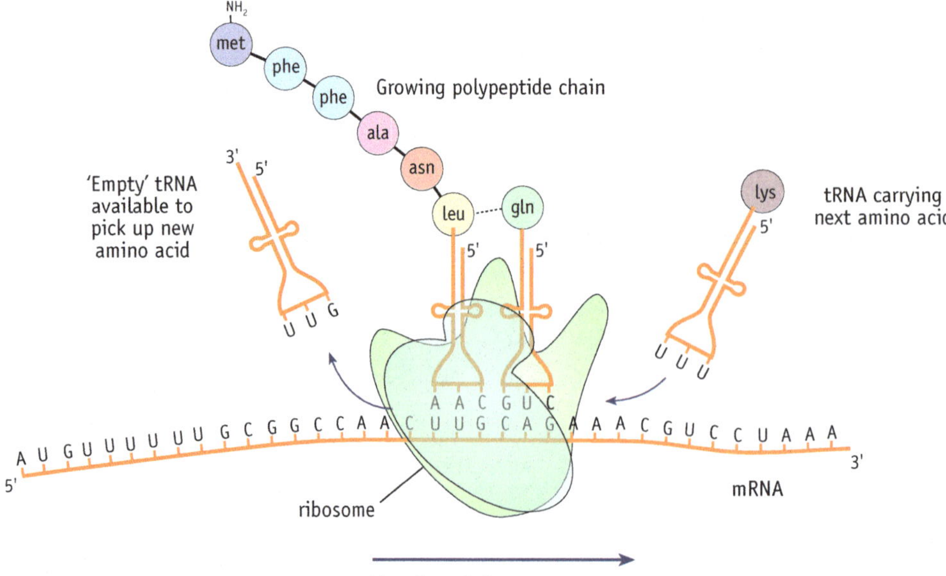

Fig. 2.7.12 A ribosome 'reading' mRNA. Each amino acid is shown by a different colour, and several have already linked to form a polypeptide. Because of the sequence of bases in the mRNA, each amino acid can only be placed in its correct position. The mRNA derived its base sequence from a gene in DNA.

7

ISBN: 9780170372855

Activity C

1 The top line gives 15 DNA bases. Reading from left to right, transcribe these into mRNA codons underneath, then into tRNA anti-codons. Finally, use the mRNA genetic code table to translate all this into amino acid names in the boxes.

DNA	GGT	GTA	TTC	ACA	CAC
mRNA	CCA	CAU			
tRNA	GGU				
amino acids	pro				

2 The middle row in the table gives seven mRNA codons. Reverse-transcribe these into DNA bases on the top line, then translate into an amino acids sequence in the boxes.

DNA							
mRNA	AUG	UGG	GGU	GGC	GGA	UCG	UGA
amino acids							

3 State where exactly in a cell each of these processes happens: transcription ____________________;

translation ____________________.

4 Write the word(s) that match each definition or description in the table below. Select from this list: *codon, transcription, tRNA, ribosomes, peptide, mRNA, translation, triplet, polymerase, anticodon.*

a	The actual sites of protein or polypeptide synthesis	
b	A group of three unpaired bases on tRNA	
c	Group of three bases on mRNA, coding for one amino acid	
d	Group of bases on DNA, complementary to a codon	
e	Its function is to carry amino acids to the ribosomes	
f	Carries information from DNA to the ribosomes	
g	Enzyme that catalyses bonds between amino acids	
h	The chemical bond between adjacent amino acids	
i	The process of DNA being copied to form RNA	
j	mRNA being used to form a particular polypeptide	

Unit 3 | Gene mutations

A **gene mutation** (aka **point mutation**) is a permanent change in any one single base in DNA. If one base is either inserted or deleted, this is not a trivial event, as it creates a major effect known as a **frameshift**.

Effects of gene mutation

For non-DNA examples, look at the following sentences.

original message:	HIS CAT ATE THE RAT	= sense
delete S:	HIC ATA TET HER AT	= nonsense
insert T at position 3:	HIT SCA TAT ETH ERA T	= nonsense
replace T with R in position 6:	HIS CAR ATE THE RAT	= mis-sense

This shows that altering a single letter can cause an extensive change in the 'downstream' message.

There are three kinds of gene mutation:

- **deleting** one base — will probably cause a frameshift, and nonsense
- **inserting** one base — will probably cause a frameshift, and nonsense
- **replacing** one base with another — unlikely to cause frameshift; could cause mis-sense.

Activity A

These questions are based on codons written in mRNA language.

1 Write the sequence of amino acids produced by the following sequence of bases

mRNA	CAG	UAU	CCC	AAG
amino acids	gln			

2 Delete the fourth base from the left of the sequence in question **1**, write out the new mutation sequence, then work out the new sequence of amino acids produced.

mutated mRNA				
amino acids				

7

ISBN: 9780170372855

3 Insert the base G after position 4 (between 4 and original 5) of the sequence in question **1**, write out the new mutation sequence, then work out the new sequence of amino acids produced.

mutated mRNA				
amino acids				

4 Substitute the base A in position three from the left of the sequence in question **1**, write out the new mutation sequence, then work out the new sequence of amino acids produced.

mutated mRNA				
amino acids				

5 State which of the three situations (deletion, insertion, substitution) produced least change in the amino acids, and explain why.

6 State which of the above three situations (deletion, insertion, substitution) produced most change in the amino acids, and explain why.

Mis-sense and nonsense

As the 'his cat ate the rat' example shows, **replacing** (substituting) one base with another is in most cases likely to cause **mis-sense** mutation; a minor spelling mistake, though sometimes with serious results (see E box, page 180). Deleting or inserting a base causes a **frameshift**, and results in a **nonsense** mutation. It may even result in a stop codon such as UAG, which would shorten the polypeptide. On the other hand, if the mutation changes a stop codon to a functional one such as UAU, then the amino acid chain will become even longer. Either way, the resulting polypeptide is likely to be non-functional. In the case of sickle cell anaemia, just one amino acid in the wrong place can have fatal results (see E box).

Mutation causes

Gene mutations may be induced by environmental agents known as **mutagens.** There are three general kinds of mutagen: certain viruses, radiation such as UV, X-rays and gamma rays, and some chemicals. There are many mutagenic chemicals, including formaldehyde and benzene compounds. This does not make mutagens automatically dangerous. Low doses of mutagenic radiation or chemicals may have no detectable effect.

7

In some cases, mutations appear to be spontaneous — which could also mean that we simply don't yet know the external cause. Mutations also occur from mistakes in DNA replication.

Cells have enzymes whose specific function it is to cut out faulty DNA sections, then replace them with the correct version. However, some mutations still get through this protective system.

Mutation characteristics

Gene mutations and also chromosome mutations (see Biology 2.5, Unit 1) have the following characteristics:

- Mutations are **rare**. Example: for achondroplastic dwarfism, the number of new mutant cases is one per 71,000 births.
- **Useful** mutations are even more rare.
- Evidence indicates that mutations are **random**, and that useful mutations don't become more likely if they are needed.
- Most mutant alleles are **recessive**, so are not expressed in heterozygotes. However, achondroplastic dwarfism, which involves a single base substitution, is caused by a dominant allele.

If a mutation is present in **gametes** (sex cells), it can be passed on to the next generation. Any mutation originating in **somatic** (body) cells cannot be passed on, though in some cases somatic mutations can cause cancers.

E

Sickle cell anaemia

This is a genetic disease resulting from a point mutation affecting one of the two genes for haemoglobin. In this situation, one single molecule of a polar amino acid (glutamic acid) is replaced by a non-polar amino acid (valine). Only one amino acid out of the 146 in each beta chain is affected.

The result of this seemingly minor change: at low oxygen concentrations, haemoglobin molecules bond to each other (instead of to water) and form crystals. This causes red blood cells to distort from their normal disc shape and become rigid, sickle-shaped cells.

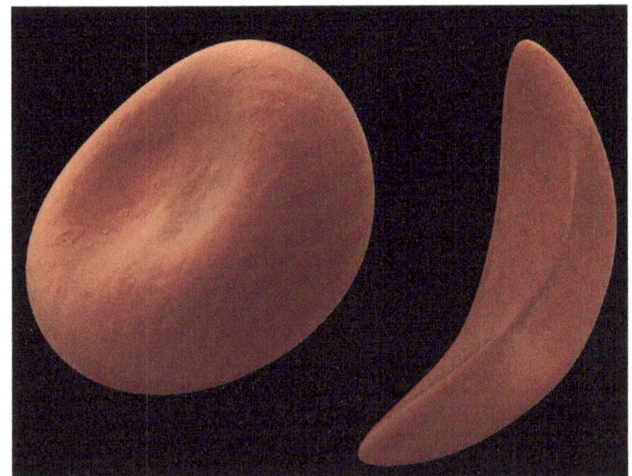

A sickle-shaped red blood cell compared with a normal one.

Babies homozygous for this allele die very young. Heterozygotes have almost no symptoms, and in fact have the advantage of being naturally resistant to malaria. This explains why the gene for sickle cell anaemia is quite common in parts of West Africa, where malaria is a major killer of those who do not carry the allele.

ISBN: 9780170372855

Activity B

1 Name two categories of frameshift mutations.

2 Explain the difference between nonsense and mis-sense mutations.

3 Name two kinds of mutagenic compounds.

4 Explain why frameshift mutations are likely to have major consequences.

5 If a mutation happens to affect only the third base in a triplet, there may be no change in the amino acid specified. Explain why this is so.

6 Explain why a mutation causing the codon UAG is likely to have an adverse effect.

7 Compare and contrast the protein synthesis stages of transcription and translation.

ISBN: 9780170372855

Unit 4 | Pathways; environment

This unit deals with two different topics:

- **Metabolic pathway**; a sequence of chemical reactions leading to an end-product.
- **Environment**; the influence that environment has on gene expression and on phenotypes.

Metabolic pathways

Metabolism is a general word for the many chemical reactions that go on in any active cell. Most metabolic processes are part of a multi-step chain of events. Each step is controlled by one enzyme, and each enzyme is the 'expression' of one or more genes.

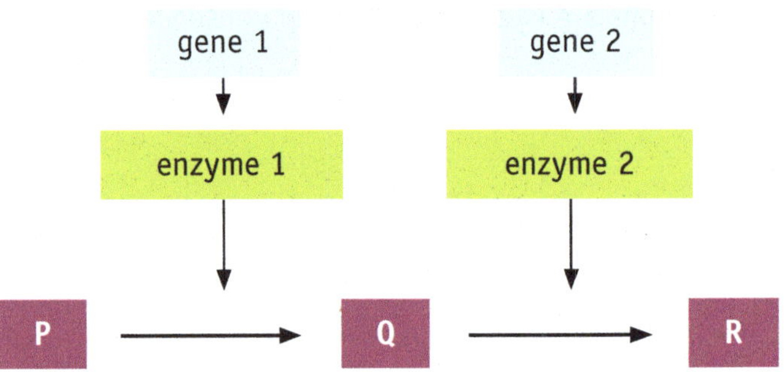

Fig. 2.7.13 In this generalised metabolic pathway, substance P is the precursor (forerunner), Q the intermediate substance, R the end product. Genes 1 and 2 and enzymes 1 and 2 all have to be functioning correctly in order for R to be produced.

Melanin pathway

A good example can be seen in the production of melanin, the dark-coloured pigment in skin and hair. We produce more melanin when we tan, in response to UV radiation in sunlight. Fig. 2.7.14 shows the amino acids phenylalanine and tyrosine, with melanin as the protein end product. If there is a defect in the gene responsible for producing the enzyme PAH (phenylalanine hydroxylase), this causes a person to develop the disease **phenylketonuria** (PKU), resulting in a build-up of phenylalanine in the body. If one of the genes responsible for tyrosinase is defective, then melanin is not produced, resulting in **albinism.** (Everyone, including PKU sufferers, also gets tyrosine direct from food.)

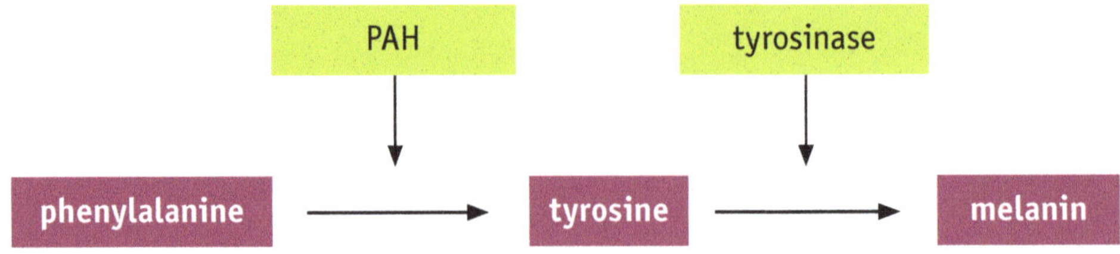

Fig. 2.7.14 Part of the metabolic pathway of melanin production, showing only one of the five enzyme-controlled steps between tyrosine and melanin.

 ISBN: 9780170372855

E

PKU and albinism

Normally there is a 'pool' of amino acids available in the blood, with a rough balance between supply (from food) and demand (mostly for protein synthesis). The gene for PAH is expressed only in liver cells. A genetic failure to produce PAH causes levels of phenylalanine in the blood to rise more than 20-fold, resulting in brain damage and mental retardation — a serious condition known as phenylketonuria, PKU. Sufferers are also very lightly pigmented due to a shortage of tyrosine and therefore of melanin.

'Melanin' is a general name for a whole group of coloured protein substances: some black, some red, some brown. If PAH is present but any of the metabolic steps between tyrosine and melanin fail to work, then the affected person will be albino, with little or no melanin in hair and skin and eyes. Albinism is inherited, an unusual condition that is commonest in black Africans. Both PKU and albinism are recessive traits.

Tyrosine is not used exclusively to make melanin — it is also an intermediate compound in the production of other substances including adrenaline. Any surplus is used as an energy source.

Flower colour example

The flowers of one variety of sweet pea are either purple or white, the final purple pigment being made in a two-step pathway under the control of two different genes. If either of these genes is defective, the purple pigment can't be made and the flowers are white. But because white can be due to a defect in either of the two genes, it is possible for two true-breeding white-flowered parents to produce purple-flowered offspring.

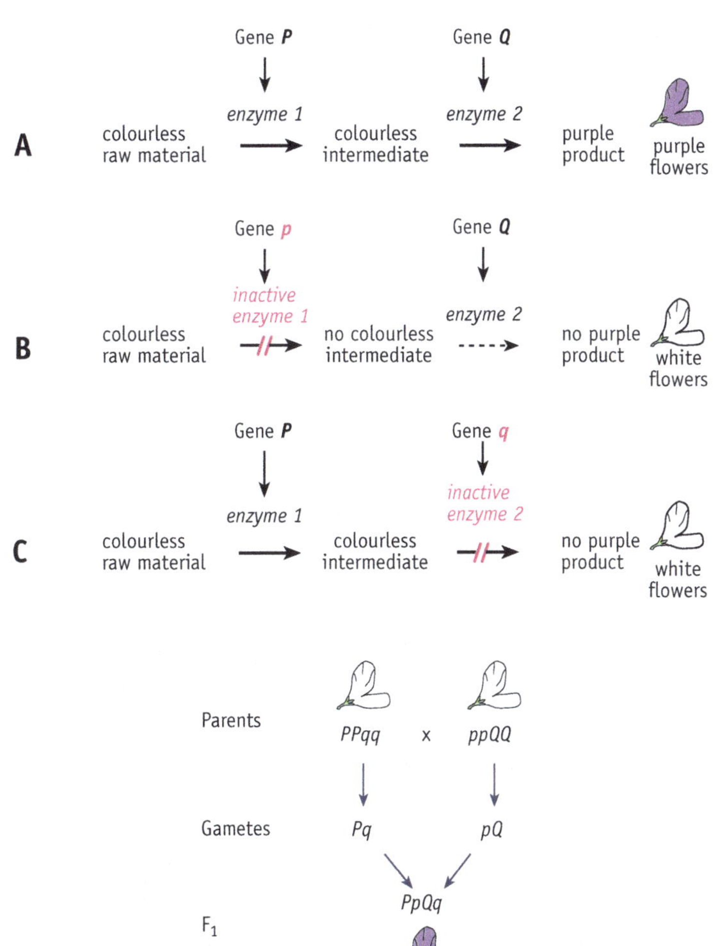

Fig. 2.7.15 A metabolic pathway in sweet peas. In purple-flowered plants, the pigment is synthesised in two steps, catalysed by two different enzymes and under the control of two different genes, shown here as *P* and *Q*. If either of these genes has homozygous recessive alleles, an enzyme will be defective and the purple colour is not produced — as shown in situations B and C.

ISBN: 9780170372855

Activity A

1 Write the word(s) that match each definition or description in the table below. Select from this list: *metabolism, melanin, tyrosinase, PKU, precursor, PAH, catalyse, tyrosine, expression, albino.*

a	A person or animal unable to produce melanin	
b	Amino acid intermediate in melanin production	
c	Substance at the beginning of a metabolic pathway	
d	A general word for chemical activity in cells	
e	One enzyme responsible for metabolising tyrosine	
f	A chemical or other result of gene action	
g	Condition resulting from excess blood phenylalanine	
h	The enzyme phenylalanine hydroxylase	
i	What enzymes do, generally	
j	Protein polymer with dark-brown colour	

2 Explain what is meant by a 'metabolic pathway'.

3 This question relates to the metabolic pathway for tyrosine and melanin, so refer to Fig. 2.7.14. Allele *J* enables production of the enzyme PAH, allele *j* does not. Allele *K* enables production of the enzyme tyrosinase, allele *k* does not. Most people have the genotype *JJKK* but there are eight other possibilities. For each of the genotypes listed below, state whether the person's 'health' phenotype will be PKU, or else normal. Also state whether their skin colour will be pale, or albino, or normal.

a *JJKK* *health* ______________; *skin* ______________

b *JJKk* *health* ______________; *skin* ______________

c *JJkk* *health* ______________; *skin* ______________

d *JjKK* *health* ______________; *skin* ______________

e *JjKk* *health* ______________; *skin* ______________

f *Jjkk* *health* ______________; *skin* ______________

g *jjKK* *health* ______________; *skin* ______________

7

ISBN: 9780170372855

h *jjKk* *health* ____________; *skin* ____________

i *jjkk* *health* ____________; *skin* ____________

j George suffered from PKU and had pale skin. His wife, Mary, had a normal phenotype. None of their children or their many descendants show any signs of PKU or albinism. Suggest the likely genotypes of George, ____________, and of Mary, ____________.

k There is no record of George's ancestors having PKU. Suggest how he inherited this condition.

____

____

4 This represents a metabolic pathway involved in the synthesis of end product Z from precursor W.

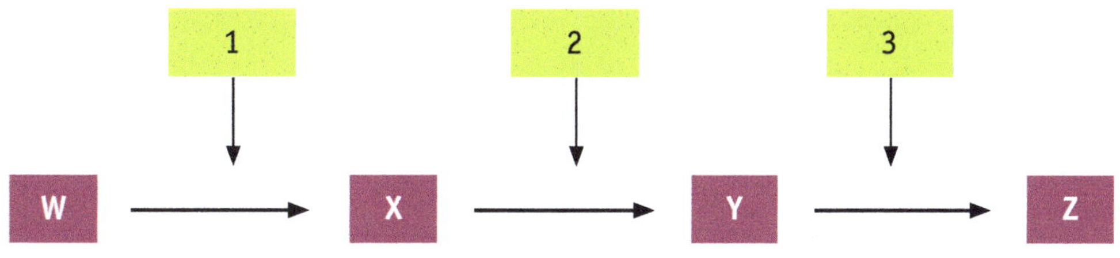

a State what numbers 1, 2, 3 stand for. ____________

b State the minimum number of genes involved in this pathway. ____________

c In terms of increase or decrease of WXYZ amounts in the body, describe the likely result of a mutation that caused substance 2 not to function.

____

____

____

d A person is suffering from a condition associated with a build-up of Y in the body. Suggest probable direct and indirect causes of this build-up.

____

____

____

____

e Suggest the source of precursor W. ____________

7

ISBN: 9780170372855

5 This question relates to the metabolic pathway for the production of purple colour in sweet pea flowers, so refer to Fig. 2.7.15. For each of the genotypes listed below, state whether the plant will have purple or white flowers.

a *PPQQ* ______________________

b *PPQq* ______________________

c *PPqq* ______________________

d *PpQQ* ______________________

e *PpQq* ______________________

f *PPqq* ______________________

g *ppQQ* ______________________

h *ppQq* ______________________

i *ppqq* ______________________

j This relates to a self-pollination of a double heterozygote F_1 plant, genotype *PpQq*. These plants will produce four kinds of gametes, whose genotypes are:

____

6 Complete the Punnett square below showing possible F_2 genotypes produced by the F_1 *PpQq* plants after they self-pollinate.

	PQ	*Pq*	*pq*	*pQ*
PQ				
Pq				
pq				
pQ				

7 From this Punnett square, predict the likely ratios of purple flowers to white flowers in the F2 generation.

____

7

ISBN: 9780170372855

What is a gene?

A gene is a unit of inheritance. Chemically, a gene is a long sequence of bases contained within DNA, but ideas have continued to develop on how best to define gene functions. One early idea was that all genes express themselves as enzymes:

one gene → one enzyme

The definition was then extended to include proteins generally:

one gene → one protein

It was then realised that several genes are needed to produce the several polypeptides that typically make up one protein, so the concept was changed to:

one gene → one polypeptide

This statement is probably true in many cases, but it now appears that in many other cases a single gene can be involved in producing different proteins in different circumstances.

Also, some genes are known to exert control over other genes in complex ways — not necessarily through polypeptide sequences — and scientists are uncertain how all this works.

In genetic modification, genes are artificially moved from one species into a different species. In some cases, these transplanted genes express themselves in exactly the same way in their new cellular 'environment'. In other cases, they do not. At present, the expression of transplanted genes cannot be reliably predicted.

Over 90% of our genome has no known function. It was formerly described as 'junk DNA' — but in reality we don't understand its role. It may be that no single definition covers all genes.

Environmental influences on gene expression

In many situations, the extent to which a gene is expressed (shows up in the phenotype) is influenced by environmental conditions. In some cases, we understand how genes and environment interact, in other cases not. 'Environment' can include factors as varied as food, temperature, and social influences.

One example exists in the fur colour of Siamese cats. In these cats, tyrosinase is abnormally heat-sensitive and is denatured even at body temperature. Result: ears, paws, face and tail are brown, while the rest of the body is creamy-coloured because skin is warmer here.

Animal examples

- In tuatara, eggs are more likely to develop into males at warmer temperatures above 22 °C and into females at temperatures below 20 °C.
- When a person is on a prolonged low-carbohydrate diet, the insulin gene becomes less active. In a person eating more carbohydrate, more insulin is produced.
- Body mass is influenced by food intake.
- In fair-skinned people, the rate of melanin production increases under the influence of UV radiation.
- Arctic foxes have white hair in winter and brown fur in summer, a gene-expression effect triggered by changes in day length — not directly by temperature.
- Siamese cat example, described above.

7

ISBN: 9780170372855

Plant examples

- Day length has a strong effect on flowering in many species. Example: 'long-day' plants produce flowers when day length (photoperiod) is longer than a critical duration. 'Short-day' plants produce flowers only when photoperiod is shorter than a critical duration. They are known respectively as 'long-day' and 'short-day' plants.
- Many kinds of tree grow vertically when surrounded by others, but sprawl outwards when growing on their own.
- Many kinds of tree and shrub show 'wind clipping', with most growth being away from the direction of strong prevailing winds.

E

Identical twins

Identical twins are monozygotic, and have been used to find the comparative importance of 'nature' (genes) and 'nurture' (environmental influence). Occasionally, identical twins are separated at birth and adopted into different families, with the result that they grow up in different environments. When comparisons are done after 30 years or more, any difference in the twins can give valuable insight into whether a particular phenotype feature is the result of nature or nurture. Examples: left-handedness, musical talent, personality tendencies, etc.

Identical twin studies are more easily done in domestic animals. If twins are deliberately separated when young and reared under different managed conditions, results can reveal much about the effect of almost any environmental factor on gene expression.

In plants, the situation is even simpler, as hundreds of cuttings (which are genetically identical) can be taken from one parent plant, then grown under different environmental conditions.

Activity B

1 Explain what is meant by 'gene expression'.

7

2 If average climate temperatures increase by 2 °C, predict what effect this could have on tuatara populations.

ISBN: 9780170372855

3 Complete each of the following sentences.

a In humans, the effect of UV light on skin is ______________________________

____

____;

and the adaptive value of this is ______________________________

____

____

____

b In arctic foxes, the effect of decreasing day length is______________________________

____

____;

and the adaptive value of this is ______________________________

____

____

4 Eight Siamese kittens from the same litter were at age 10 weeks separated into two groups of four, and both groups then kept under artificial indoor conditions for six months. Group 1 kittens were kept at temperatures 25 °C to 30 °C, while group 2 kittens were kept at temperatures 5 °C to 10 °C. Both groups were equally well treated, given all the food they needed, and the same kinds of food.

a Predict the likely appearance of each group of kittens after two weeks of separation. Justify your prediction.

____

____

____

b Predict the likely appearance of each group of kittens after six months of separation. Justify your prediction.

____

____

____

____

____

7

ISBN: 9780170372855

5 Some hydrangea plants produce blue flowers, some produce red. The difference could be due to genes, or could be due to soil type.

a If cuttings taken from a blue-flowered plant and grown in a variety of soils eventually all turn out blue-flowered, explain what conclusion can be drawn.

b If cuttings taken from a blue-flower plant and grown in a variety of soils eventually produce some blue-flower plants and some red-flower ones, explain what conclusions can be drawn.

6 Explain the mechanisms that cause Siamese cats to have cream-coloured bodies and dark brown extremities.

 ISBN: 9780170372855

Biology 2.7 Gene expression

AS 91159 Demonstrate understanding of gene expression
Externally assessed, 4 credits

Achievement	Achievement with Merit	Achievement with Excellence
Demonstrate understanding of gene expression.	Demonstrate in-depth understanding of gene expression.	Demonstrate comprehensive understanding of gene expression.

Achievement
'Demonstrate understanding ...' involves defining, using annotated diagrams or models to explain, and giving characteristics of gene expression, or an account of it.

Achievement with Merit
'Demonstrate in-depth understanding ...' involves providing a reason as to how or why biological ideas and processes affect gene expression.

Achievement with Excellence
'Demonstrate comprehensive understanding ...' involves linking biological ideas and processes about gene expression. The explanations may involve justifying, relating, evaluating, comparing and contrasting, or analysing.

Notes
(Summary below. For further details on 2 and 4, visit the NCEA website.)

1 'Gene expression' involves a selection from the following biological ideas and processes:
- nucleic acid structure and nature of the genetic code
- significance of proteins
- protein synthesis
- the determination of phenotype via metabolic pathways
- effect of environment on genotype through mutations
- effect of environment on expression of phenotype.

2 Biological ideas and processes relating to nucleic acid structure and nature of the genetic code are selected from:
- molecular components and their role in carrying the genetic code
- nature of the genetic code including triplets, codons and anticodons
- redundancy due to degeneracy within the code.

3 Biological ideas and processes relating to the significance of proteins are selected from:
- proteins as the products of gene expression: DNA → mRNA → polypeptide or protein
- identification of one gene → one polypeptide relationship
- significance of proteins is limited to their structural and catalytic roles

4 Biological ideas and processes relating to protein synthesis are selected from:
- the role of DNA sequence in determining the structure of a protein
- the role of enzymes in controlling the process (specific enzymes not required).

5 Biological ideas and processes relating to the determination of phenotype via metabolic pathways are selected from:
- biochemical reactions are catalysed by specific enzymes and every enzyme is coded for by a specific gene(s)
- biochemical reactions do not occur in isolation but form part of a chain reaction so that the product of one becomes the substrate of another step in metabolism
- phenotype is determined by the presence, absence, or amount of specific metabolic products.

6 Biological ideas and processes relating to the effect of the environment on genotype through mutations are selected from:
- mutagens (specific mutagens are recognised but their effect at molecular level is not required)
- the potential effect on genotype and phenotype of gene mutations at the gene level.

7 Biological ideas and processes relating to the effect of environment on expression of phenotype involve ways that environmental factors may change phenotype without changing genotype.

Revision 5 | Gene expression

Write a list of key points in each box.

Proteins

DNA and coding

Metabolic pathways

Genes and environment

ISBN: 9780170372855

Revision 6 | Gene expression crossword

Across

6 Chain of amino acids (11)
10 Weak bonds holding protein molecules in shape (8)
11 Sub-units of proteins (5, 4)
12 Describes a sequence of three nucleotides (7)
13 Structure of a globular protein consisting of more than one polypeptide (10)
14 Mutation resulting from a frameshift (8)
19 Mutation that causes one base to be replaced by another (8)
20 Collagen is an example of this kind of protein (7)
21 Describes the shape of enzyme molecules (8)
22 One product of condensation reactions (5)
23 Any molecule consisting of many smaller units linked in a chain (7)

Down

1 Process in which a strand of DNA is copied (13)
2 Genetic code with some amino acids represented by more than one triplet (10)
3 Guanine's relationship to cytosine (13)
4 The pattern of folding in globular protein (8)
5 Exact place where translation occurs (8)
7 Mutation resulting from insertion or deletion of a base (10)
8 Sequence of amino acids in a polypeptide (7)
9 Splitting of a large molecule with the addition of water (10)
12 Molecule that brings amino acids to mRNA (8, 3)
15 Describes mutations that occur in body cells (7)
16 An RNA copy of the gene (9, 3)
17 Any mutation involving a single DNA base (5)
18 Group of atoms with acid properties, in amino acid (8)
19 Any agent that causes mutations (7)

7

ISBN: 9780170372855

Exam-type question 1

DNA and RNA are nucleic acids that carry the genetic information responsible for the synthesis of proteins in the cell.

- Compare and contrast the structure of DNA with RNA.
- Relate their structure to how they are able to code for the production of proteins.
- Explain what is meant by 'degeneracy of the code'.

Use this scaffold framework as the basis for an answer to be done on your own paper.

- Compare chemical similarities and differences of DNA and RNA.
- For example, 'Both are made of … with a "backbone" of … and … However, the sugar components in DNA are …, whereas in RNA they are … Both DNA and RNA have four kinds of bases, but in RNA … in place of the … in DNA.'
- Contrast. How many strands do DNA and RNA have? How long are they? Describe the general task/function of each; both the similarities (= coding) and the differences.
- Now relate DNA/RNA structure to their coding function, as follows:
- Define 'gene'. Where are genes 'held'?
- Describe how a gene is copied into a messenger molecule. Name this molecule.
- Explain why DNA 'needs' this intermediate messenger.
- Explain what is meant by 'transcription'.
- Describe how DNA determines/controls where the gene to be transcribed begins and ends. Mention coding strand, template, promoter sequence, coding region, terminator sequence.
- Name the RNA-controlled process that occurs in the cytoplasm.
- Name the structures involved. Name the end product(s) of this process.
- Describe how a messenger molecule indicates where to start and stop translation. Describe the start and stop codons of the messenger molecule.
- Explain what 'degeneracy' of the code means.
- State how many amino acids are coded for. Example: 'A two-base system of codons could code for only … possibilities. A three-base system can code for …, which is more than is needed.'
- Explain why in the mRNA table there are two or more codons for each amino acid.

 ISBN: 9780170372855

Exam-type question 2

Protein synthesis is a two-step process involving 'transcription' in the nucleus, followed by 'translation' in the cytoplasm. Complementary base pairing is critical to ensuring correct gene expression.

Discuss this statement addressing the following points in your answer:

- Describe the possible complementary base pair combinations possible in transcription.
- Explain what causes the base pairs to match up in this way.
- Explain how the coding and template strands of DNA relate to mRNA.
- Explain how triplets, codons and anti-codons relate to the process of translation of a strand of mRNA.

Use this page to create a key points plan or a mind map or scaffold that could form the basis for a longer answer to be done on your own paper.

ISBN: 9780170372855

Exam-type question 3

Clover plants have a metabolic pathway in which two genes are required to produce the enzymes necessary for the production of hydrogen cyanide. This chemical is toxic to grazing herbivores, and plants that possess both the dominant alleles needed to produce it are said to be cyanogenic.

Study the metabolic pathway shown and discuss it, ensuring that you address the following points in your answer:

Gene ***A*** → *Enzyme 1*; Gene ***B*** → *Enzyme 2*

inactive precursor —(*Enzyme 1*)→ cyanogenic glycoside —(*Enzyme 2*)→ cyanide + glucose

- A definition of metabolic pathway.
- Using the information in the flow diagram, explain how the ability of the plant to produce hydrogen cyanide depends on its genotype.
- Explain how the cyanogenic plants may have arisen in the population.
- Explain the reason for the persistence in the population of the cyanogenic clover plant.

Use this scaffold framework as the basis for an answer to be done on your own paper.

- Define 'metabolic pathway'.
- Using *A* as the symbol for enzyme 1 production and *B* for enzyme 2 production, explain how the presence or lack of a dominant allele determines whether the pathway is completed.
- Explain why *aaB–* and *A–bb* genotypes produce no cyanide.
- Explain why *A–B–* is the only genotype that can produce cyanide.
- Explain the source (origin) of new alleles in a population, referring to hydrogen cyanide in your answer.
- Explain how the alleles *A* and *B* persist in a population by detailing any adaptive advantage producing this chemical could give to a plant.
- Relate your answer to the process of natural selection.

7

ISBN: 9780170372855

Exam-type question 4

Cretinism and albinism are two disorders caused by breakdown in a particular metabolic pathway in humans. In the homozygous condition the affected individuals are unable to produce the critical enzymes required.

Study the diagram and answer the question that follows.

Key:

1 Cretinism (characterised by retarded growth, especially of the brain) results from lack of thyroxine.

2 Albinism results from lack of melanin (skin pigment).

Phenylalanine and tyrosine are amino acids.
E = various enzymes (names not required)

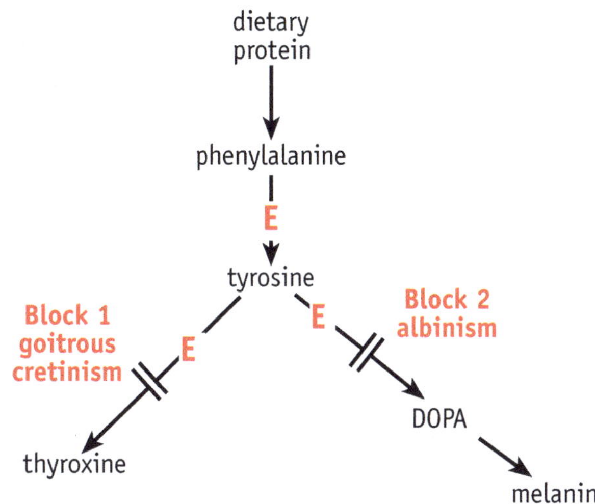

Discuss the metabolic pathway outlined above referring to the diagram and include:

- three ways in which a gene mutation can result in a non-functional enzyme.
- the role of enzymes in causing cretinism and albinism.
- an explanation of how two people could suffer from albinism resulting from mutations in different genes.

Use this page to create a key points plan or a mind map or scaffold that could form the basis for a longer answer to be done on your own paper.

ISBN: 9780170372855

7

Exam-type question 5

Use the mRNA table on page 175 to help you answer this question.

Consider this sequence of bases on the template strand of DNA coding for a polypeptide:

TAC GAG **ATA** ACG TTA ACC

Use your knowledge of transcription and translation to discuss the effects that point mutations can have on the amino sequence being coded for.

- Describe the original mRNA sequence coded for by the DNA strand.
- Describe the amino acid sequence resulting from this code.
- Describe the types of point mutations — substitution, insertion and deletion.
- Explain the possible effects of these mutations if they were to occur at the third triplet on the template above (shown in **bold**).
- Use examples from the mRNA table to help illustrate your meaning.

Use this scaffold framework as the basis for an answer to be done on your own paper.

- Define mutation.
- If this 'SC' mutation is harmful, is it likely to be a deletion, addition or substitution mutation? (Read the stimulus material carefully.)
- Explain why some mutations are more likely (than others) to cause an effect. Use these terms: frameshift, amino acids, functional polypeptide.
- Explain how bonds hold polypeptides in shape. Mention tertiary structure and how this is critical to the shape of red blood cells.
- Explain how the shape of red blood cells relates to their function of carrying oxygen.
- Explain how a mutation could cause this change in shape.
- Link to how this would affect blood cells' function.
- Explain how sickle cell anaemia still occurs today, despite it being harmful.
- Link your explanation to the process of natural selection.

7

ISBN: 9780170372855

Exam-type question 6

Some metabolic pathways are complex. One particular multi-step pathway produces melanin as its end product. Albinism is a condition that results if the individual does not produce any melanin in their skin and hair. Below is a simplified representation of the metabolic pathway producing melanin.

substrate --------> (gene A) intermediate substrate --------> (gene B) coat colour

Using this flow chart and your knowledge of metabolic pathways, discuss how albinism may be caused in an individual. In your discussion, include the roles of:

- DNA
- enzymes
- gene mutation
- environment.

Use this page to create a key points plan or a mind map or scaffold that could form the basis for a longer answer to be done on your own paper.

ISBN: 9780170372855

Exam-type question 7

Siamese cats have an interesting colour pattern called colour pointing, in which only the extremities (nose, ears, paws and tail) of the cat are chocolate coloured, whereas the rest of the cat is cream or white (albino). To produce the pigment melanin an enzyme tyrosinase is required. This converts tyrosine to melanin through an intermediate called DOPA. In Siamese cats a mis-sense mutation arose in the gene coding for the protein tyrosinase, so that it is abnormally temperature sensitive. The kittens are born totally albino but at around four months old start to develop the colour-pointed pattern.

tyrosine $\xrightarrow{\textit{tyrosinase}}$ DOPA $\xrightarrow{e}$ melanin (chocolate pigment)

e = enzyme

Discuss colour pointing in Siamese cats by addressing the following:

- How the environment and genotype contribute to the expression of colour pointing.
- Explain how temperature could affect the activity of the mutant form of the enzyme tyrosinase in the nose, ears, paws and tail.
- Give a reason for why the colour-pointed pattern only develops in kittens at around four months.

Use this scaffold framework as the basis for an answer to be done on your own paper.

- Describe the genotype for a colour-pointed cat.
- Explain how the metabolic pathway works step by step, relating to the two genes and their protein products. Example: 'Gene A produces the enzyme ..., which converts ... to ...'
- Explain how temperature can affect the structure of a protein. Include: stability of its shape; tertiary structure; bonding.
- Explain why temperature differs in different parts of a cat's skin.
- Explain why the body of the cat is not affected and remains pale in colour, while extremities turn chocolate as the metabolic pathway can be completed.
- Consider conditions in the womb pre-birth. After being born, which parts of the kitten probably cool the most? Relate this to the metabolic pathway and colour production.
- This change is not immediate. Consider how long fur takes to grow.

 ISBN: 9780170372855

Exam-type question 8

Refer to the mRNA table on page 175 of this workbook to answer the following question.

Part of a DNA strand coding for an enzyme has the following base sequence:

Coding strand	TAC	GGC	CTC	CCG	TGC	AAT	CAG	ATA	TGG
Template strand	ATG	CCG	GAG	GGC	ACG	TTA	GTC	TAT	ACC
mRNA strand	___	___	___	___	___	___	___	___	___
Amino acids	___	___	___	___	___	___	___	___	___

Complete the table above and then discuss these results by addressing the following:

- Explain how the third base could change in any one of the triplets without causing a change in the amino acid produced.
- Explain how you can tell from where in the particular gene this sequence is taken.
- Explain the significance of the codons AUG and UAG.
- Why is the code based on a three-letter (triplet), not a two-base system?
- What is meant by saying that the code is 'universal'?

Use this page to create a key points plan or a mind map or scaffold that could form the basis for a longer answer to be done on your own paper.

Exam-type question 9

'An organism's phenotype results from a complex interaction of its environment with its genotype.'

Discuss this statement using relevant plant and animal examples to support your explanation of how environment can contribute to causing phenotypic differences in organisms having the same genotype.

Use this scaffold framework as the basis for an answer to be done on your own paper.

- Recall at least two plants and two animals that you have studied. Describe in each case how environment changes the appearance, despite having the same genotype.
- Example: 'Manuka trees growing together in large numbers tend to grow tall and straight because … However, the growth shape of manuka growing singly in exposed coastal areas tends to be …'
- Relate the difference(s) to environmental factor(s), considering which in the case of manuka is the most likely cause. Soil stability? Soil moisture? Wind? Light?
- Other examples named in this book: insulin production; human skin colour; human body mass; tuatara; photoperiod in flowers; arctic foxes; honeybees.
- For each example, make sure you deal with the one organism in detail explaining why the difference is evident, fully relating phenotype to environmental differences.
- Choose examples that you can easily remember and describe.

7

ISBN: 9780170372855

Exam-type question 10

Tuatara are endemic reptiles with an interesting method of sex determination in offspring. If eggs are buried in cool soil on south-facing slopes with temperatures below 21 °C they develop into females, while those incubated at warmer temperatures develop into males. This is referred to as temperature-sensitive sex determination or TSD. It appears that the enzyme aromatase (which converts testosterone into estrogen) is important in temperature-dependent sex determination.

In sea turtles the reverse is true. Males develop in cooler temperatures of 22.5 – 27 °C and females in warmer temperatures.

This basis for sex determination has been used by some scientists to hypothesise how the dinosaurs may have become extinct after asteroids struck the earth.

Discuss how environment affects sex determination in reptiles including:

- the role of enzymes in this metabolic pathway,
- an explanation of how a knowledge of TSD could help in the conservation of endangered species such as tuatara and sea turtles,
- an explanation of how scientists think that TSD could have contributed to the extinction of many of the dinosaurs.

Use this page to create a key points plan or a mind map or scaffold that could form the basis for a longer answer to be done on your own paper.

ISBN: 9780170372855

Microscope skills

NCEA Achievement Standard 91160

Investigate biological material at the microscopic level

2.8

Internally assessed, 3 credits.

Unit 1 | Microscope types and uses

This standard develops hands-on skills and can be done together with externally assessed Achievement Standard 91156 (Biology 2.4). Standard 2.8 can be graded as Achievement or Merit, but not Excellence. It involves:

- using a microscope correctly to get good-quality images
- preparing your own slides: two different plant tissues, one microscopic animal
- making three or more annotated drawings that match your prepared slides. Identify special features in cells you have drawn. Give reasons how and why these features enable these cells to carry out specific functions.

You will be assessed on the quality of your three prepared slides, and also on your three or more annotated drawings based on the slides.

Types

There are four general types of microscopes, the first two often used in schools:

- binocular microscope (aka stereo microscope)
- student microscope (aka compound microscope)
- microscope with computer screen display
- electron microscope (EM). These enlarge a million times or more and cost around $1 million. It's possible your school does not have one of these, yet.

A binocular microscope, also known as a stereo microscope. Usual magnification: 10x to 40x. This type is suitable for solid objects and requires lighting from above. (Alternative: a hand lens can be used as a simple microscope.)

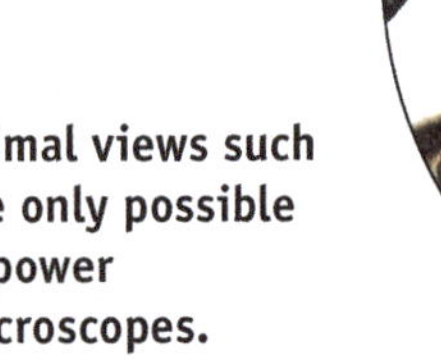

Whole-animal views such as this are only possible with low-power stereo-microscopes.

ISBN: 9780170372855

Microscope skills

Activity A shows an ordinary student microscope. This type is suitable only for very thin transparent material and requires lighting from below. The usual magnifications are 40x, 100x, 400x. Some models have two eyepiece lenses.

The following steps apply mainly to student microscopes. These need to be done in sequence, and will help you see objects more clearly. Special care is needed at step 6.

1 Set objective lens to low power (LP).
2 Place slide on stage; hold it in place with clips.
3 Focus gently with the coarse focus control; move slide to target area.
4 Change to medium power (MP). Re-centre slide if needed. Refocus using fine control.
5 Change to high power (HP) only if necessary. Do this with great care while looking from the side. Reason: to prevent the lens touching the slide. When looking through the eyepiece, adjust fine focus with very small movements.
6 Change back to MP or LP before removing the slide.

At any point, you can improve image sharpness by closing the diaphragm slightly and adjusting the mirror (if any). Generally, a darker background and smaller diaphragm will give sharper images.

The **magnification** ('mag.') of a microscope is easily calculated:

objective lens mag. x eyepiece lens mag. = total combined magnification

It is also useful to know the **field of view**: the diameter of the circle you see.

Activity A

Label parts **1** to **11** on the photograph. Select from this list: *coarse focus control, base, high-power objective lens, low-power objective lens, eyepiece lens, stage, stage clips, condenser lens and diaphragm, fine focus control, revolving turret, mirror.*

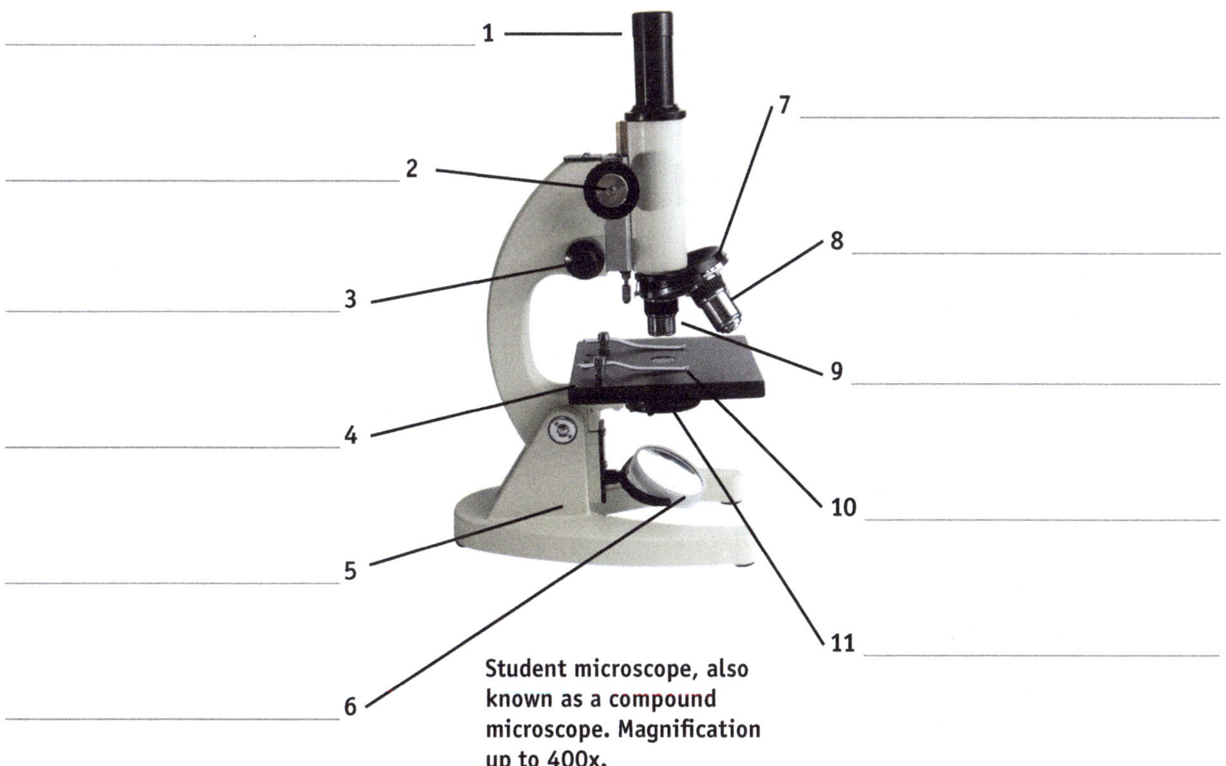

Student microscope, also known as a compound microscope. Magnification up to 400x.

ISBN: 9780170372855

Activity B

1 The table is missing the first half of each sentence. Write in the blank spaces to make seven correct complete sentences, choosing from these words: *The diaphragm is used for, An objective lens is, Only the fine focus control, The turret is, The stage is, Stage clips are used for, An eyepiece is.*

	Part	Function or description
a		is to be used with high-power lenses.
b		the rotating part holding three or four lenses.
c		the lens closest to your eye.
d		the one closest to the slide.
e		holding the slide in place.
f		regulating the amount of light.
g		the platform on which slides are placed.

2 Compete the table by writing in the 'Reason' column.

	Rule	Reason
a	When not in use, always put a plastic cover over a microscope.	
b	If the microscope has a mirror, never use it to reflect sunlight onto the slide.	
c	Always look from the side when moving the HP objective lens slowly into place.	
d	Never let water or your fingers or any object touch a lens.	
e	Never force a focus control.	

 ISBN: 9780170372855

Activity C

This involves using a student microscope.

1 Calculate the overall magnification of your microscope under

LP: __________ (eyepiece mag.) x __________ (objective mag.) = __________________

MP: __________________ x __________________ = __________________

HP: __________________ x __________________ = __________________

2 Place the edge of plastic ruler on the stage and measure the field of view at LP and MP, i.e. find the width of the 'image circle'.

LP: __________________ mm

MP: __________________ mm

HP: (The field at 400x will be ¼ as wide as at 100x) ______________________ mm

3 Write a general rule about the link between magnification and how wide the field is.

____

____

4 When looking from the side, estimate the clearance between the objective lens and slide for:

LP: __________________ mm

MP: __________________ mm

HP: __________________ mm (This is the reason you need to be so careful with HP.)

5 Cell sizes are measured in micrometres, symbol μm.

1 μm = 10^{-6} m (one millionth of a metre)
= 10^{-3} mm (one thousandth of a millimetre)

Complete this table.

Object	Size in mm	Size in μm
Average diameter of a human red blood cell	7/1000 mm	
Diameter of HP field for a student microscope	0.4 mm	
Diameter of a leaf epidermis cell	0.1 mm	
Typical diameter of a chloroplast	1/100 mm	
Typical length of a mitochondrion	0.003 mm	
Typical length of *Paramecium* cell		200

8

With student microscopes, it's not possible to see detail smaller than about 10 µm.

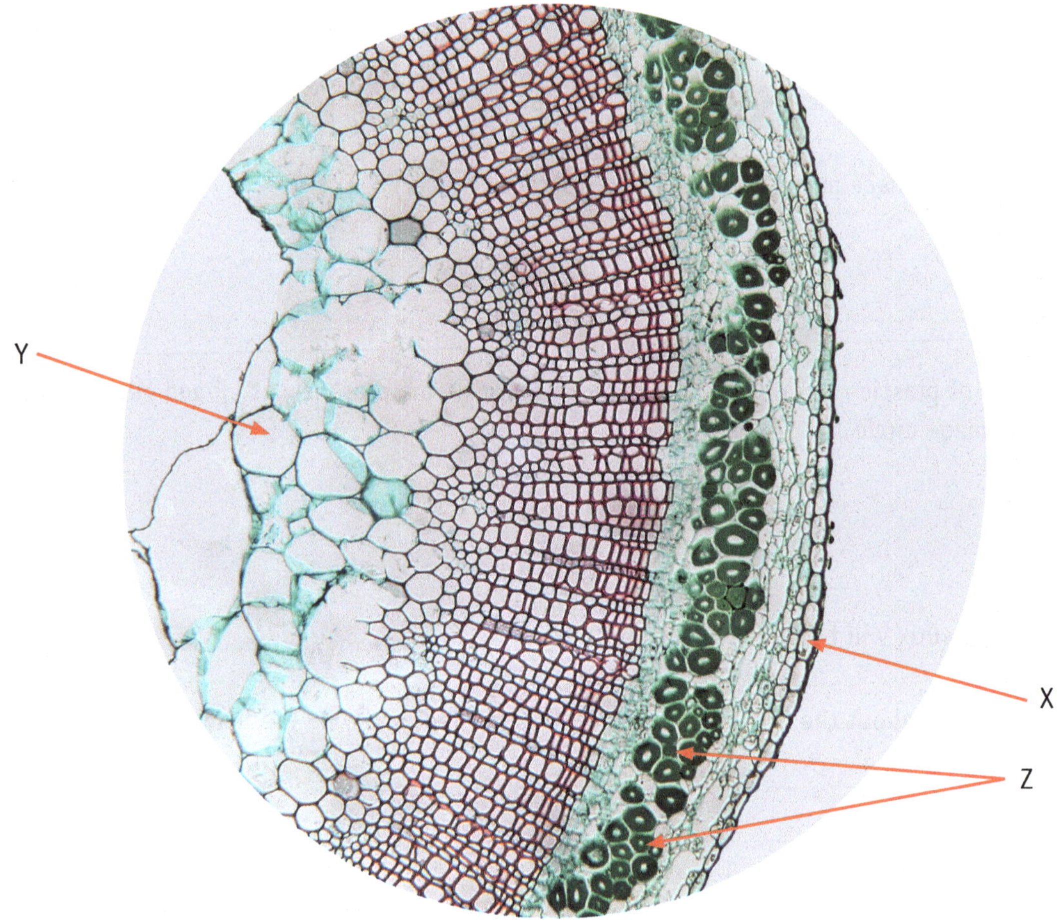

6 The photograph shows a cross-section of a plant stem. As shown here, the enlargement is 200 times.

a Measure the diameter of the circle: _____________ mm. This means that the actual field of vision in this photograph is _____________ mm, which is the same as _____________ µm.

b Measure the length of cell X in the photo: _____________ mm.

So actual length = _______ µm.

c Measure the diameter of cell Y in the photo: about _____________ mm.

So actual diameter = _______ µm.

d Describe one distinctive feature of the cells marked Z.

____

____

e Suggest a possible special function of the cells marked Z.

____

____

8

ISBN: 9780170372855

Unit 2 | Slides and drawings

Slide basics

In most situations, the object to be looked at is placed on a flat glass slide. Any living or 'fresh' material needs to be immersed in water to prevent the cells from drying. Problem: each drop of water acts like a distorting lens. Solution: use a **coverslip** to put a flat 'lid' over the water drop. A coverslip always needs to be lowered very carefully, as shown in Fig. 2.8.1. Reason: to avoid trapping air bubbles that can create further problems.

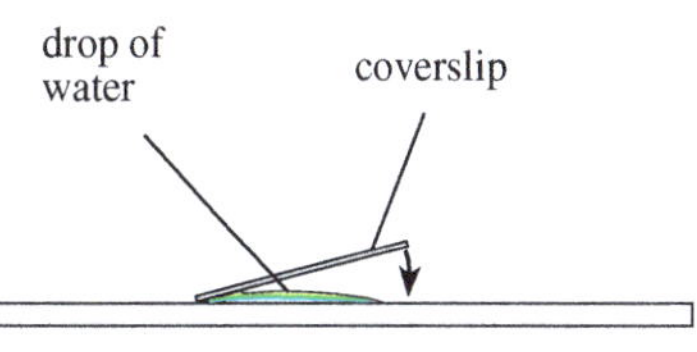

Fig. 2.8.1 Side view of coverslip being lowered.

- If too much water is used in the initial drop on the slide, some of it can be removed by gently touching a piece of filter paper to one edge of the coverslip.
- If you are looking at live swimming microscopic organisms such as *Paramecium*, see the section 'Techniques for protists' (page 210). In these situations it may be better to use a **cavity slide** — but without a coverslip. With a cavity slide, use only LP and MP, never HP.
- Slides can be cleaned and reused, but coverslips are too fragile to be cleaned. Slides are cheap; coverslips more expensive; cavity slides even more.

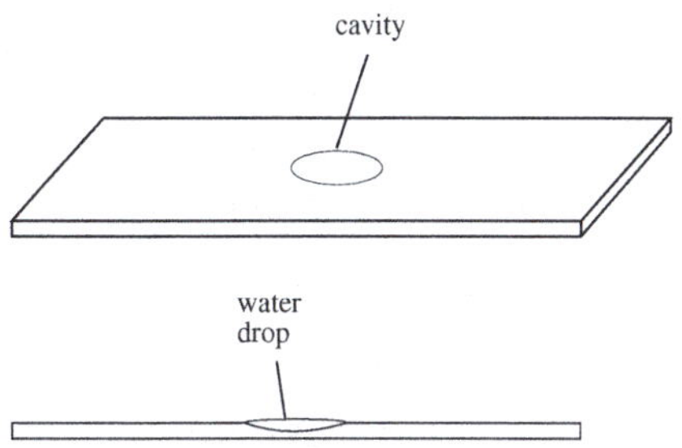

Fig. 2.8.2 Cavity slide.

Staining

Thin layers of cells are almost colourless. This makes it hard to see details, so we use chemical stains to make cells and their contents more visible. **Iodine solution** is a stain often used for plant cells. Iodine gives a blue-black colour when it encounters starch, including starch granules inside chloroplasts. Other parts of the cell become stained yellow-brown. Add one drop of stain to one side of the coverslip, and pull it towards the plant or animal material by briefly touching a piece of filter paper to water on the opposite side of the coverslip.

Toluidine blue is another stain often used. Most stains kill cells, which means that if you want to look at live microscopic animals you will need to use other methods to see them clearly. The usual techniques involve methyl cellulose (see below) or carefully readjusting the lighting.

Preparing plant slides

Even if coverslip and stains are correctly used, it is difficult to see cell details if the tissue is many cells thick. Use methods A and B to get thin layers only one or two cells deep.

A Epidermis peels. Use a sharp spike or pair of forceps to peel a very thin layer of skin from a leaf. Before it has a chance to dry out, immediately place it in a drop of water on a slide. This takes practice, even though you need a patch of epidermis only a few millimetres in diameter. All plants have an epidermis, but smooth soft leaves are best. Two good garden plants for making peels with specialised cells are *Kniphofia* (red hot poker) and *Chlorophytum* (spider plant). Take epidermis peels from the underside of leaves, as these have more stomata. Onion peels are not suitable because their cells are not specialised.

ISBN: 9780170372855

B Leaf sections. Place a freshly picked leaf in a slit cut into a piece of raw potato or carrot (to hold the leaf in place). Use a very sharp new razor blade to slice the leaf crossways; only a few millimetres length. Immediately place the leaf slice (section) in a drop of water on a slide before it has a chance to dry out. It may take lots of practice before you can make a slice thin enough. *Griselinea* (papauma) is suitable, but many kinds of leaf will do. For **stem sections**, use the leaf-slicing technique with stems from soft plants like tomato. Sections can be cut lengthways (LS) or crossways (CS). Both stems and leaves have many kinds of specialised cell.

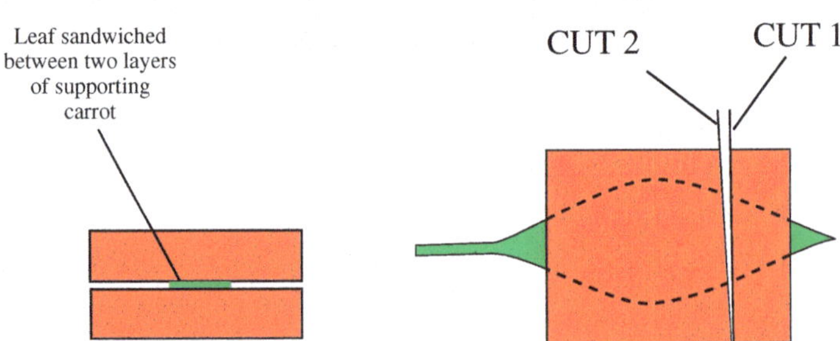

Fig. 2.8.3 Cutting a thin section by hand. If you cut the stem or leaf section at a slight angle, its tapered end may be only one or two cells 'deep'.

For epidermis peels and leaf sections and stem sections, the next steps are:

- have a preliminary look under LP to see if the section is thin enough
- if it's worth keeping, add a coverslip
- stain with iodine solution
- observe, draw.

Techniques for protists

One-celled protists such as *Paramecium* have specialised organelles. Examples: cilia, food vacuoles plus contractile vacuoles for osmo-regulation. If *Paramecium* is not available, there are many other suitable protists including *Stentor* and *Chaos*. All of these can be found in non-polluted ponds with decaying plant matter. If you wish to keep a long-lasting supply of protists, use the internet for advice on culture-growing techniques. Start months before you need a culture.

If you are looking at microscopic organisms that move too fast to be seen properly, they will usually slow down if you add a wisp of cotton wool to their drop of water. Alternatively, prepare a viscous mixture that will slow them down. Mix 1.5 g methyl cellulose powder into 50 mL boiling water. Cool, then stir 50 mL cold water into the mixture. The mixture should last few weeks. Add a drop of this solution to the water the animals are swimming in. This usually causes them to swim slowly, but their behaviour may become unnatural, with contractile vacuoles becoming very big.

***Stentor* are trumpet-shaped protists that live in pond water.**

8

ISBN: 9780170372855

Activity A

1 Complete this table on specialised cell features in plant leaves and stems. Select from this list: CO_2, *water loss, xylem, photosynthesis, water, sunlight*. (One of these is used twice.)

Cell or tissue feature	Function
Many chloroplasts	Site of photosynthesis; absorbing as much ____________ as possible
Epidermis cells transparent	To enable more light to enter leaf interior, to maximise ____________
Stomata	Guard cells open to assist ____________ entry for ____________
Epidermis cells thick-walled	To protect against insect and wind damage, also to reduce ____________
Long thick-walled cells with openings at the ends	For carrying ____________, as in the case of ____________ cells

2 Write the matching word(s) in each blank space below. Select from this list: *coverslip, epidermis, iodine, micrometre, methyl cellulose, toluidine, cavity slide, chloroplasts.*

a	A solution that stains starch blue-black	
b	Special slide that can contain several drops of water	
c	Thin flat piece of glass for covering a specimen	
d	Blue stain suitable for some plant material	
e	Green organelles inside many kinds of plant cell	
f	Cells on the outer layer of a plant or leaf	
g	A solution of this can be used to slow 'swimmers'	
h	Unit of length equal to one thousandth of a millimetre	

ISBN: 9780170372855

Quality drawings

You do not have to be artistic to create high-quality biology drawings. The skills lie in your eyes, and in using these six points as a guide:

1. Look! Draw what you see, not what you think should be there.
2. Make the drawing *big*; at least half a page.
3. Use a sharp pencil, not a pen.
4. Draw clean lines; not fuzzy-edged sketchy lines, not shaded.
5. Draw a few parts carefully, not many parts in a hurry. Three cells drawn with details carefully observed are better than 50 cells drawn carelessly. To show fine detail, inset drawings can be helpful. Ignore unimportant details such as air bubbles.
6. Use indicator lines to identify and name important features. These lines should not cross over and the writing should be kept away from the drawing itself.

Around your drawing add further information (annotations) such as:

- Title.
- Stain used.
- Microscope magnification used to view the cells. (This is *not* the same as the amount of enlargement from reality to drawing.)
- A scale line, e.g. 0.1 mm (100 μm). Remember: HP field ≈ 0.4 mm wide; MP ≈ 1.6 mm.
- Cell colour, cell shape, visible cell contents, cell wall thickness.
- In the case of active life forms, describe their movements, including contractile vacuole.
- Information on how each of these details (shape, cell contents, wall thickness, movements, etc.) is related to the **function** of that tissue. If your annotations cannot all fit around the drawing, then continue on a separate page.

You will be assessed on the quality of your three prepared slides, and on your portfolio of annotated drawings based on these slides.

Tissue: a number of similar cells specialised for a particular function.

ISBN: 9780170372855

Activity B

Below are two students' drawings of plant leaf cross-sections. One is graded as 'Not achieved', the other as 'Merit'. Add labels that comment on errors of detail in each drawing. Also state what important items of information are missing from each. Add annotations that might improve the 'Merit' drawing. (Note there is no 'Excellence' grading for 2.8.)

1 'Not achieved'

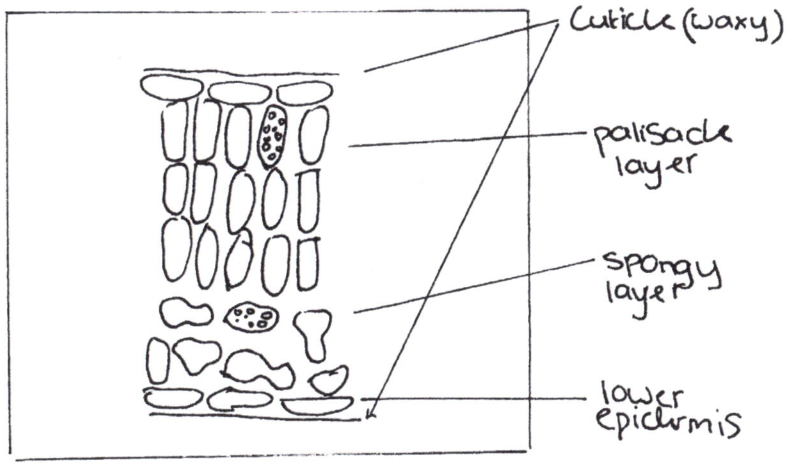

2 Merit

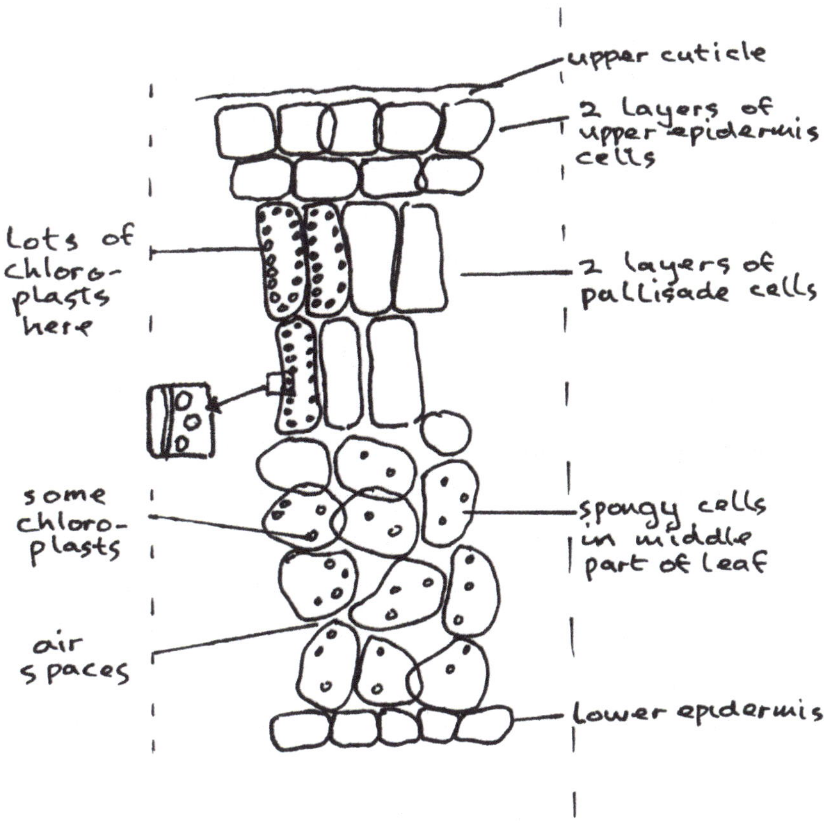

ISBN: 9780170372855

Biology 2.8 Microscope skills

AS 91160 Investigate biological material at the microscopic level
Internally assessed, 3 credits
Note: This standard is not graded at Excellence level.

Achievement	Achievement with Merit
Investigate biological material at the microscopic level.	Investigate in-depth biological material at the microscopic level.

Achievement
'Investigate ...' involves:
- preparing biological material for viewing under a light microscope
- viewing biological material using a light microscope to enable detail of cell structures and components to be determined
- recording observations of biological material in biological drawings
- identifying observed specialised features and relating them to the function of the cells or tissues.

Achievement with Merit
'Investigate in-depth ...' involves:
- giving reasons how or why observed specialised features enable the cells to effectively carry out their specific function(s).

Notes
1 'Biological material for viewing' includes two different plant tissues and one unicellular organism.
2 To allow an accurate drawing to be produced, preparation of material may include: staining, use of cavity slides, use of cellulose, epidermal tear, cutting sections.
3 A biological drawing follows the accepted conventions to record observations consistent with the biological material being viewed. Consistency must include recognisable shape and proportions and inclusion of typical organelles present in a cell, appropriate to the magnification. At the Achieved grade, the biological drawing may contain some errors in applying conventions or minor inaccuracies in representation. At the Merit grade, the biological drawing may contain some minor errors as long as they do not affect the accuracy of the representation of the biological material being viewed.
4 Specialised features may include: arrangement of cells or cell types within a tissue, shape of a cell, presence or absence of a specific organelle, quantity or distribution of organelles within a cell. Notes about the specialised features may accompany the biological drawing (e.g. a fully annotated diagram).
5 Relating observed specialised features to the function of the cell or tissue must include: identifying the feature or organelle, stating its function, and giving reasons for why or how it contributes to the function of the cell or tissue.

ISBN: 9780170372855

Glossary

Allele Alternative forms of a gene. Alleles occur in pairs, usually with two alleles per gene
Alveolus (plural: alveoli) One of millions of minute, blind endings of the air passages in the mammalian lung
Anticodon Sequence of three unpaired bases on tRNA, complementary to a codon on mRNA
Autosomes Chromosomes not concerned with the determination of sex; in mammals and *Drosophila*, all chromosomes apart from X and Y

Centriole Part of animal (and some plant) cells that organises the spindle in cell division
Centromere Part of a chromosome that attaches to spindle fibres during cell division
Chloroplast Organelle in which photosynthesis occurs
Chromatid One of two identical products of replication of a chromosome, joined at the centromere
Chromosome Long sequence of DNA, consisting of many genes, that behaves as a single unit during mitotic cell division
Coding region The main part of a gene, containing the base sequence required to make a polypeptide
Codon Any sequence of three bases in mRNA that specifies one particular kind of amino acid
Controlled variables in an experiment, these are all the variables that could possibly affect results, and need to be kept as constant (controlled) as possible
Crossing over Exchange of sections of non-sister chromatids during the first meiotic division
Cytosis Process in which substances enter or leave a cell by 'pinching off' small areas of plasma membrane

Dependent variable In an experiment, this is the factor likely to depend on or be influenced by the independent variable. The dependent variable gives the results (data)
Diffusion The movement of particles (atoms, ions, molecules) of any substance from where they are more concentrated to where they are less concentrated
Dihybrid Product of a cross between true-breeding organisms differing with respect to two characteristics
Diploid Having two sets of chromosomes

Enzyme Biological catalyst that increases the rate of a particular reaction, or in some cases reduces the rate
Epithelium Layer of cells forming a protective covering or the lining of a cavity
Eukaryote Organism in which the chromosomes are enclosed in a distinct nucleus

Family tree A diagram showing family genetic relationships, including that of siblings

Gamete Sex cells; eggs and sperm. Gametes cannot develop until fertilisation happens
Gas exchange The two-way diffusion of oxygen and carbon dioxide across a surface
Gene A unit of inheritance, consisting of a length of DNA that codes for one particular kind of polypeptide
Gene mutation A change in the base sequence of a gene
Gene pool All the genes in a population
Genetic drift Random process in which allele frequencies are changed by chance events
Genotype The genetic information carried by an organism
Glycolysis Process in the cytoplasm in which six-carbon glucose is partly broken down into three-carbon pyruvate
Golgi complex (aka **Golgi apparatus**, **Golgi body**, **Golgi**) Cytoplasmic organelle in which newly synthesised protein molecules are 'packaged' before they are sent to their destination

Haploid Having a single set of chromosomes
Heterozygous Having different alleles of a particular gene

Homologous chromosomes Chromosomes that pair during meiosis and cannot normally be present in the same gamete
Homozygous Having two copies of the same allele
Hypothesis A suggestion or intelligent guess that can be tested in some way

Independent variable In an experiment, this is the factor that is deliberately chosen to be varied as part of the experiment's aim

Krebs cycle Aerobic process in mitochondria in which organic molecules are oxidised to CO_2 and useful energy

Lamella A structure with a flat leaf-like shape, increasing surface area
Lux A physical unit of light intensity, symbol lx. Direct sunlight is 30,000 to 100,000 lx
Lysosome Cytoplasmic organelle containing digestive enzymes

Meiosis Process in which a diploid nucleus divides twice to form four haploid, genetically different nuclei. Halves the number of chromosomes per nucleus, and in animals, produces gametes
Meristem Region of a plant in which cells are constantly dividing (e.g. root and stem tips)
Mitochondrion Organelle in which aerobic respiration occurs
Mitosis Cell process in which a nucleus divides to form two genetically identical nuclei
Monohybrid Product of a cross between true-breeding organisms differing with respect to a single characteristic
Mutagen Any agent that causes mutations; which can be caused by some chemicals, radiation or viruses

Nucleolus Region of the nucleus concerned with production of ribosomal RNA
Nucleotide Sub-unit of nucleic acid, consisting of a five-carbon sugar, a phosphate group, and one of five kinds of nitrogenous base

Okazaki fragments Short, initially separate, sections of the 'lagging strand' produced during DNA replication

Organelle Structurally distinct part of a cell specialised for a particular function
Osmosis Movement of water across a semi-permeable membrane from higher to lower water potential

Pedigree A diagram representing an individuals' direct male and female ancestors, but not siblings
Phenotype The bodily characteristics of an organism
Plasma membrane Membrane forming the outermost layer of the cytoplasm
Plasmolysis Process in which the cytoplasm of a plant cell shrinks away from the cell wall as a result of loss of water to an external solution of lower water potential
Polyploid Having more than two sets of chromosomes
Proboscis In insects, a specialised tube-like mouthpart used for sucking up fluids
Prokaryote Organism in which the chromosomes are not enclosed in a distinct nucleus
Promoter region Section of DNA at the start of a gene, instructing RNA polymerase where to begin transcription

Quadrat A sample part of a habitat used in ecological surveys. Quadrats can be large or small

Respiration (cell respiration) Biochemical energy-releasing process that occurs in most cells

Siblings Brothers and sisters
Stratification Vertical layering of vegetation in a forest
Succession Any process of community change that happens over many years, often following a disturbance such as fire or erosion
Synthesis Any process that involves 'putting things together' to create something greater, such as the joining of smaller molecules

Template strand DNA strand used to synthesise a complementary DNA strand
Terminator region Section of DNA at the end of a gene, instructing RNA polymerase where to stop transcription

Zonation Changes in plant and animal species present along an environmental gradient in a habitat
Zygote A fertilised egg

ISBN: 9780170372855

Answers

2.1 | Practical investigation

Unit 1

Activity A (page 8)

1 a If we grow tomato seedlings in soils with a range of pH 4 to 10, then they will grow fastest at pH 7 to 8.
  b If we give them a range of food choices, slaters will prefer damp wood.
  c Below a certain concentration, the disinfectant will be ineffective.

2

a	independent variable (IV)
b	dependent variable (DV)
c	controlled variables (CVs)
d	valid
e	data
f	hypothesis
g	questions
h	conclusion
i	table
j	prediction

Activity B (page 10)

1 The same individual tree at the same time of the year.
2 Remove single small branches and measure all the leaves, not just some.
3 Perhaps 50–100 leaves in each sample.
4 Perhaps maximum width.
5 Measure light intensities with a meter. Measure height above ground with a tape. Use compass to check exactly where north and south are.
6 One person might pick leaves/branches at a higher level than the other; or might unconsciously choose bigger/shinier/healthier looking leaves.

Unit 2

Activity A (page 12)

1 One hypothesis: perhaps oily foods release more energy than other kinds of food.
2 IV: different kinds of food.
3 At least four different kinds.
4 The temperature rise of water in the flask.
5 Degrees Celsius.
6 Same weight of food to be burned, equally dry, started burning in the same way, same amount of water, same-sized container, same starting temperature, same thermometer is used in the same way.
7 Major defects: most of the heat from the flame will be lost and not go into the water; some heat will then be lost from the water. Partial answers: have a wider-base flask, and insulate the sides. Use a metal container since metal conducts heat.

Activity B (page 14)

1

Light (lx)	Bubbles per minute	Average
250	5, 4, 5, 3, 6	4.6
500	21, 15, 17, 23, 18	18.8
1200	43, 49, 59, <u>450</u>, 49	50.0
2000	62, 47, 57, 59, 61	57.2
3000	59, 70, 64, 62, 62	63.4
3500	69, 66, 60, 59, 62	63.2

2 Discard the '450' result, as it is far outside the other results. It may have arisen from a decimal point error.
3 (Graph)
4 About 60 to 63 bubbles per minute
5 About 150 to 200 lx
6 Independent variable: light brightness, range 250 to 5000 lx
7 Variables that would need to be controlled: same amount of plant, type of plant, temperature, concentration of carbon dioxide, time of day.
8 The results show that at low light intensities the rate of photosynthesis increases in proportion to light intensity. Under bright conditions the rate of photosynthesis levels off and reaches a limit. This is probably caused by CO_2 availability becoming a limiting factor.

Activity C (page 16)

1 IV: concentration of the salt solution
2 IV units: % salt, or else mol/L
3 DV: weight change of the kumara pieces
4 DV units: gain or loss of weight as a % of starting weight
5 CVs: same-diameter pieces, same length, cut from the same kumara, same-sized containers, same volume of solution, same balance used for weighing, same length of time, same amount of blotting before final weighing
6 Three pieces per solution x 5 solutions = 15 total
7 Put each piece in its numbered solution immediately after weighing.
8 Weigh each piece immediately after cutting, and later immediately after blotting.

Activity D (page 19)

1 An enzyme is a biological catalyst.
2 (10), 20, 30, 40, 50, 60, (70) degrees Celsius
3 Add hot water to the water bath; icy water for lower temperatures.

2.2 | Analysing information

Unit 1

Activity A (page 24)

1	evidence
2	proof
3	hypothesis
4	valid
5	theory
6	analyse
7	bias
8	objective
9	vested interest

Activity B (page 26)

Answers will vary.

2.3 | Plant and animal adaptations

Unit 1

Activity A (page 34)

Cactus

1 Spikes deter grazing animals, so less likely to be eaten
2 Having no leaves will reduce transpiration, so more likely to survive drought
3 Flowers best at times of water abundance, which is also when pollinating insects are available, so more likely to produce seeds
4 Waxy surface reduces transpiration, so more likely to survive drought
5 Epidermis almost transparent: allows maximum light into the stem so more photosynthesis and faster growth
6 Stem swells after rain and stores water for dry times, so more likely to survive drought

Mosquito

1 Antenna helps locate food (you), essential for its niche as a bloodsucking parasite
2 Big eyes, good eyesight: help locate food and avoid predators, so survives longer
3 Proboscis reaches their food supply (blood beneath the skin), so obtains food quickly
4 Saliva prevents clotting so it can get more blood food in a short time
5 Flexible body so can eat a big meal in a short time, so survives longer
6 Stomach enzymes able to digest protein: to digest its blood meal

Activity B (page 36)

1

a	adaptation
b	biochemical
c	niche
d	taxon
e	organism
f	transpiration
g	gas exchange
h	epidermis
i	genetic
j	proboscis

2 a Beak strong, has a sharp hooked curve
b Adaptive advantage: ripping up prey
c Silent flight: can swoop on prey undetected, since at night-time mice, etc. more likely to rely on hearing than vision
d Judging distances accurately important as owls rely on this to catch prey at first hit
e Ruru have big eyes, big pupils
f Big eyes: this enables them to see in low light, which makes for successful hunting

Unit 2

Activity A (page 39)

1 Gas exchange surfaces tend to have a big surface area because diffusion is a very slow process, so a large area increases total uptake.
2 Gas exchange surfaces need to be kept moist because diffusion in solids is very slow, slightly faster in liquids, so a wet surface is better than dry skin.
3 Gas exchange surfaces like lungs tend to lose water rapidly because these surfaces have to be kept moist.
4 Diffusion can be defined as the movement of atoms, ions or molecules of a substance from where they are more concentrated to where they are less concentrated.

Activity B (page 40)

The following fish structures and functions could be noted. Each point should relate to specific drawing feature(s), preferably using indicator lines, and should explain *how* that feature makes gas exchange faster or more efficient.

- gill filaments and gill lamellae greatly increase total surface area for gas exchange
- gill rakers reduce the chances of lumps of food getting tangled up with and damaging the gill filaments
- abundant blood supply, which increases the concentration gradient of gases
- water flow across the gill surface brings a supply of O_2, removes CO_2
- water flow and blood flow are in opposite directions (a counter-current system), which increases the concentration gradient
- gill surface very thin; epithelium only one cell thick

Activity C (page 41)

The following mammal structures and functions could be noted. Each point should relate to specific drawing feature(s), preferably using indicator lines, and should explain *how* that feature makes gas exchange faster or more efficient.

- millions of microscopic alveoli greatly increase total surface area for gas exchange
- epiglottis helps prevent food going down the wrong way and getting into the lungs
- abundant blood supply, which increases the concentration gradient of gases
- movements of the diaphragm and the intercostal muscles bring a supply of O_2, remove CO_2,which increases the concentration gradient of gases, thus speeding up the rate of gas exchange
- lung surface is very thin, with apparently only one or two epithelial cells between air and blood
- pleural membranes lubricate the surface of the lungs as they move, thus reducing friction.

Activity D (page 43)

The following insect structures and functions could be noted. Each point should relate to specific drawing feature(s), preferably using indicator lines, and should explain *how* that feature makes gas exchange faster or more efficient.

- tracheae and tracheoles penetrate every part of the body, which greatly increases the total surface area for gas exchange
- tracheoles are in direct contact with body cells for gas exchange, so there is no need to have blood transporting O_2 and CO_2
- air sacs together with body movements help pump air in and out of the trachea system, which brings a supply of O_2, removes CO_2, which increases the concentration gradient of gases
- tracheole 'walls' are very thin, only one cell thick, which increases the concentration gradient
- spiracles can close to reduce evaporation and water loss, and perhaps prevent dust, etc. from entering.

Activity E (page 44)

1

a	spiracle
b	tracheole
c	capillary
d	diaphragm
e	rakers
f	filaments
g	pleural
h	epiglottis
i	pulmonary
j	larynx
k	diffusion
l	respiration

2 Insect tracheae are very narrow tubes. Oxygen diffuses inwards, CO_2 outwards. Diffusion is a slow process, made slower by different molecules moving in opposite directions. Trachea work well over short distances, but are inefficient over distances of more than a few millimetres. This is one reason why wide-body insects tend to be slow, and bigger ones don't exist.

2.4 | Cell biology

Unit 1

Activity A (page 47)

1

	Organelle or cell part	Function or main activity: brief summary
a	Cell wall (in plants only)	To protect and support the cell
b	Vacuole (large in plants only)	Fluid pressure supports the cell; also sugar storage
c	Plasmodesmata (in plants only)	Help movement of substances between cells
d	Chloroplasts (in plants only)	Photosynthesis: absorb light, produce sugars
e	Mitochondria	Respiration: produce ATP
f	Plasma membrane	Regulates entry/exit of substances into and out of cell
g	Ribosomes	Site of protein synthesis
h	Nucleus	Encloses chromosomes
i	Nucleolus	Site of RNA production
j	Endoplasmic reticulum	Transport system within the cell
k	Golgi complex	Holds secretory substances, especially modifies proteins
l	Centriole (in animal cells)	Produces spindle during mitosis and meiosis

2

Structure/organelle	In plant cells?	In animal cells?
Nucleus	Yes	Yes
Mitochondria	Yes	Yes
Ribosomes	Yes	Yes
Endoplasmic reticulum	Yes	Yes
Centriole	In a few (ferns)	Yes
Golgi complex	Yes	Yes
Plasma membrane(s)	Yes	Yes
Chloroplast	Yes (in leaves, etc.)	Never
Cell wall	Yes	Never
Large vacuole	Yes (in most)	Never

Activity B (page 50)

1 Mitochondria, chloroplasts, nucleus, Golgi complex, vacuoles, ER
2 'Semi-permeable' means that small molecules can pass through freely, but bigger ones cannot
3 Water, oxygen, CO_2
4 a Endoplasmic reticulum
b RER has ribosomes, which appear as small particles on ER surfaces; SER does not have these.
5 Ribosomal and transfer RNA
6 Vesicles containing newly produced proteins are budded off the ER cisternae and fuse with the Golgi apparatus where the proteins undergo further change.
7 Chromosomes are in the nucleus (except in bacteria).

ISBN: 9780170372855

Activity C (page 53)

1 a The primary function of chloroplasts is to convert light to chemical energy in photosynthesis.
  b The primary function of mitochondria is to make the most of a cell's ATP in respiration.
  c One similarity between the internal structure of chloroplasts and mitochondria is they both have a large surface area of internal membranes.
  d The difference between 'chloroplast' and 'chlorophyll' is a chloroplast is an organelle, which contains the light-absorbing pigment chlorophyll.
  e A simple word equation for cell respiration is organic compounds + oxygen → carbon dioxide + water.
  f A simple word equation for photosynthesis is water + carbon dioxide + light → sugar + oxygen.
  g The main function of lysosomes is to digest items inside the cell, such as worn-out organelles and phagocytosed bacteria.
  h The energy-holder known as ATP is 'recharged' from ADP and phosphate, using energy obtained from the oxidation of organic matter in respiration.
  i A function of cilia and flagella in micro-organisms is to create movement by pushing against the surrounding water.
  j The main difference between prokaryote and eukaryote cells is prokaryotes have no distinct nucleus.
  k The likely reason why mitochondria are long and thin is it gives a greater surface area compared with the volume.

2

	Energy source?	Gas used?	Gas made?	Organic molecules
Photosynthesis	light	CO_2	O_2	made
Respiration	chemical	O_2	CO_2	broken down

3 a Most mitochondria: heart muscle because adapted for contraction; liver cells because energy needed for metabolism/synthesis.
  b Least mitochondria: bone cells because little growth, little chemical activity, no movement; fat cells because they store chemical energy, but do not need much energy/ATP. Note: red blood cells have no mitochondria, as little chemical activity.

4 (a, b, c) pyruvate, ADP, phosphate (d, e) CO_2, ATP

5 (a, c) CO_2, H_2O (b) light (d, e) O_2, sugars

6 cell wall: layer of cellulose outside the cell membrane
plasma membrane: regulates movement of substances in and out of cell
microvilli: very small folded projections on surface of cells
chloroplast: photosynthesis
mitochondrion: rod shaped, up to 4 micro- metres long, folded inner membrane
nucleus: contains the chromosomes which hold coded information on protein manufacture
EPR: thin tubes that extend throughout the cell
lysosome: small vesicles containing enzymes
ribosome: sight of protein synthesis
cilia: used for swimming in small organisms, used for propelling fluids in larger animals
Golgi: for retention and further processing of synthesised proteins

7 Oxygen enters mitochondrion by diffusion, originally supplied via the bloodstream. Pyruvate enters by diffusion, originally from the process of glycolysis. ADP and phosphate enter by diffusion from elsewhere in the cell. Main factor affecting the rate of transport of each of these: a cell's level of metabolic activity and hence demand for ATP from the mitochondria.

8 *Sunlight input*: radiated into the chloroplast, after passing through surrounding leaf cells. *Water input*: reaches the chloroplast via xylem cells in the leaf, and via xylem in the stem, and originally from soil water. *Main organic product*: glucose. *Destinations of derived substances*: making new cells in leaf and stem; used in respiration; used as one of several kinds of sugars in flowers and fruits etc.

Unit 2

Activity A (page 58)

1 a nuclear envelope
  b endoplasmic reticulum
  c nucleolus
  d mitochondrion

2 about 0.25 μm

3 30mm

4 30 mm / 0.00025 mm = 120,000 mag

Activity B (page 59)

1 smooth endoplasmic reticulum
2 rough endoplasmic reticulum
3 ribosomes
4 mitochondria outer membrane
5 mitochondrial crista
6 pinocytosis
7 nucleolus
8 nuclear membrane
9 DNA/chromosome
10 mitochondrion
11 centriole
12 Golgi apparatus

Activity C (page 60)

	Small cube	Medium	Big
Length of cell	1 cm	2 cm	3 cm
Total surface area	6 x (1 x 1) = 6 cm^2	24 cm^2	54 cm^2
Total volume	1 x 1 x 1 = 1 cm^3	8 cm^3	27 cm^3
SA:V ratio	6:1	3:1	2:1

Conclusions:

1 Diffusion caused NaOH to penetrate about the same distance into each cube. NaOH reached all of the small cell's interior, but reached only a small proportion of the large cell's interior.
2 Small objects have a *large* surface area-to-volume ratio, compared with *bigger* objects of the same shape.
3 Biggest surface area: *big cell*. Absorbed NaOH to greatest extent: *small cell*.
4 No. The original hypothesis is not supported.
5 Answers will depend on results. Discussion statements will vary.

Unit 3

Activity A (page 64)

1 Diffusion, osmosis, facilitated diffusion
2 'Passive' because they don't require energy from ATP
3 Higher temperatures, steeper concentration gradient, smaller particles
4 At higher temperatures, particles move faster/have more kinetic energy.
5 Active transport and cytosis
6 ATP supplies chemical energy in a form that cell processes can utilise almost instantly.
7 Transport proteins extend through plasma membranes and have an exclusively transport function. 'Specific' means these proteins only convey particular types of molecule.
8 'Uphill' refers to movement from a region of high concentration for a particular kind of particle to a region of lower concentration.
9 a diffusion b osmosis c facilitated diffusion
  d active transport e downhill f phagocytosis
10 A (phospho)lipid layer, B protein, C CO_2, D O_2, E glucose, F protein

Activity B (page 67)

1 When a protein is denatured, this breaks the bonds that hold the secondary and tertiary folded structures in shape, and the whole molecule then permanently refolds into a different shape.
2 Protein molecules are fragile because their complex folded shapes are held in place by weak hydrogen bonds. These H bonds are easily broken.

3

a	catalyst
b	enzyme
c	active site
d	substrate
e	product
f	metabolism
g	anabolic
h	optimal
i	denatured
j	pepsin
k	induced fit
l	catabolic

Unit 4

Activity A (page 71)

1

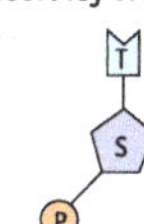

2 Base pairing rule: adenine pairs with thymine and guanine pairs with cytosine.
3 A polymer is a large molecule consisting of many smaller, similar ones linked in a chain.
4 Nucleotides
5 The function of DNA replication is to ensure that the 'daughter' DNA molecules are identical to the 'parent' molecule.
6 DNA replication is 'semi-conservative': 'semi' means partial and 'conservative' means keeping. This is because only half of each daughter DNA molecule is new; the other has been 'kept' from the parent DNA molecule.
7 DNA strand: TTG AGC ATA CCG
  Complementary: AAC TCG TAT GGC
8 C A D F E B
9 base, nucleotide, DNA double helix, chromosome, nucleus, cell
10 Complementary base pairing gives DNA a highly accurate way of copying itself and its information.
11 Okazaki fragments are short sections of DNA which form on one strand, then become linked later. They formed this way because nucleotides can only be added at a 3' end. The other side of DNA can grow continuously in this direction.

Activity B (page 74)

1 A human embryo starts off as a single cell, host to produce billions of cells within a short time (by birth), so each cell has to divide many times.
2 Mitosis is especially rapid in skin cells and also in gut lining, because these cells are continually being worn away and need to be replaced.
3 When a chromosome is duplicated, the two duplicates are known as chromatids.
4 Centrioles are the structures that form spindle fibres during meiosis and mitosis. Centromeres are the points where two chromatids are attached.
5 Daughter cells are produced by the division of a parent cell.
6 a 46 b 92 c 46 d 46
7 F I B H C G A J D E

ISBN: 9780170372855

Unit 5

Activity A (page 77)

1. Fat cell. Feature: large lipid droplet. Main function is energy storage; large to store large amounts of fat. Feature: few mitochondria. Main functions energy storage; little need to release energy or produce ATP.
2. Secretory cell in salivary gland. Feature: many RER and Golgi complexes. Main function is to produce protein on ribosomes (on RER); then to secrete proteins via Golgi. Feature: many mitochondria. Protein synthesis is anabolic so needs energy, so needs ATP, which is mainly released in mitochondria. Feature: several vacuoles. To retain protein before secretion.
3. Red blood cell. Feature: very small size. Small size means larger surface area so rapid oxygen loading and unloading. Feature: filled with haemoglobin. To transport oxygen. Feature: no organelles. These cells are specialised for oxygen transport and nothing else.
4. Epithelial cells lining an artery. Feature: thin flats cells. So as not to impede blood flow. Feature: bound tightly together. To prevent cells being dislodged.

Activity B (page 79)

1. Feature: thick cell walls. Very thick/lignified. For support (wood).
2. Feature: long thin shape. Increased surface area, to absorb water from surrounding soil.
3. Feature: shape suited for transport. Wide diameter. Feature: no end walls. To provide continuous tubes for water transport. Cell walls very thick/lignified, so they don't collapse when water is sucked through them.

Activity C (page 81)

1. Epidermis cells have no chloroplasts: to allow maximum amount of light to reach leaf interior of leaf.
2. Palisade cells at 90° to surface: to allow maximum amount of light to reach leaf interior.
3. Chloroplasts abundant in palisade cells: photosynthesis here.
4. Thousands of stomata: to speed up CO_2 absorption (and O_2 release).
5. Leaf mesophyll air spaces: to speed up CO_2 absorption for photosynthesis.
6. Vein xylem cells: supply water for photosynthesis and other activities.
7. Thick-walled cells surrounding veins: to keep leaf supported and flat and angled towards light.
8. As well as being limited by availability of light, the photosynthesis and rate is limited by other factors as well, such as enzyme activity and CO_2.
9. a CO_2 is a raw material for photosynthesis, so photosynthesis rates increase when more CO_2 is available.
   b The rate of photosynthesis also depends on other factors such as light intensity and and the rate at which enzymes can act. One or other of these will act as a limiting factor, causing the graph to level off no matter how much CO_2 is available.

Revision 2

Cell biology crossword (page 85)

Across: 1 hydrogen bond; 5 chromatid; 6 pyrimidine; 8 nuclear envelope; 11 cytosol; 12 ATP; 13 chloroplast; 14 organelle; 19 DNA; 24 osmosis; 25 micrometre; 26 plasma membrane; 27 respiration; 28 Golgi apparatus.
Down: 2 nucleus; 3 ribosome; 4 endoplasmic reticulum; 7 denaturation; 8 nanometre; 9 chromosome; 10 cilium; 15 anabolism; 16 diffusion; 17 glycolysis; 18 vacuole; 20 crista; 21 enzyme; 22 centromere; 23 centriole.

2.5 | Genetic variation and change

Unit 1

Activity A (page 97)

1. Continuous variation: a full range of in-betweens. Examples: height, weight. Discontinuous variation: traits that exist in only a few or even two forms. Examples: tongue rolling, fur colours within a particular dog breed.
2. Gene: a length of DNA coding for a polypeptide. Locus: the position of a gene on a chromosome. Allele: one form of a gene.
3. A trait is a distinctive and inherited feature of an organism, differing sharply from other such traits.
4.

Feature	Genes?	Environment?
Tongue rolling	10/10	0/10
Musical ability	7?	3?
Adult height	8?	2?
Eye colour	10	0
Milk yield in cows	5?	5?
Your accent	0?	10?
Five toes on each foot	10	0

Activity B (page 100)

1.

a	variation
b	gene
c	allele
d	locus
e	continuous variation
f	discontinuous variation
g	chromosome

h	trait
i	mutation
j	expression
k	silent
l	polyploid
m	point
n	diploid

2. Mutations and sex (including meiosis and fertilisation)
3. 1st duplication; 2nd inversion; 3rd translocation; 4th deletion.
4. Down syndrome (trisomy 21) results from having three copies of chromosome 21, instead of the normal two, so total of 47. Triploidy: three full chromosome sets.
5. Existing alleles have all proved their worth through natural selection, so any random change is unlikely to be an improvement.
6. Genetic variation makes evolution possible. Selection can only operate if there is variation, and affects future generations if this variation is to some extent genetic.

Unit 2

Activity A (page 106)

1.

a	gamete
b	zygote
c	fertilisation
d	diploid
e	haploid
f	somatic
g	crossing over
h	centromere
i	meiosis
j	mitosis

2. a In meiosis, the number of chromosomes is halved because this prevents a doubling of chromosomes that would otherwise happen at every generation.
   b Bacteria rely on mutations to create variety because bacteria do not have meiosis to create variety.
   c The difference between centromere and chiasma is a centromere is a point at which sister chromatids attach, while a chiasma is the point of crossing over between two non-sister chromatids.
   d The importance of independent assortment and crossing over is that both increase the amount of genetic variety, an essential feature of evolution.
3.

	Mitosis	Meiosis
Occurs where (in animals)?	any growth tissue	ovary/testis only
Resulting cells haploid or diploid?	diploid	haploid
Resulting cells genetically identical?	yes	no
Chromosome number?	unchanged	reduced

4. a Meiosis I resembles mitosis: two nuclei are formed (each with a full set of chromosomes or chromatids)
   b Difference: in meiosis the two nuclei are different, in mitosis they are identical.
5. a 23 **b** 46 **c** 23 **d** 23 **e** 92
6. a Since all the banana trees in a plantation are genetically the same, a disease that affects one plant is likely to affect all.
   b Mutations will occasionally create genetic variation in banana plants.

Activity B (page 109)

1. At the end of meiosis, the four nuclei drawn should each have one full-length chromatid. One all red. One all blue. One top part red, lower part blue. One top part blue, lower part red.
2. **a** *AD* **b** *AD*, *ad* **c** *Ad*, *ad*
3. **a** *AD* **b** *AD*, *Ad*, *aD*, *ad* **c** *ad*
4. **a** *AKE* **b** *AKE*, *AKe*, *AkE*, *Ake*, *aKE*, *aKe*, *akE*, *ake* **c** *AKE*, *AKe*, *AkE*, *Ake*

Unit 3

Activity A (page 113)

1.

a	genotype
b	phenotype
c	recessive
d	alleles
e	dominant
f	monohybrid

g	F_1
h	F_2
i	true breeding
j	testcross
k	discontinuous
l	locus

2. Both parents are *bb*. No. No *B* allele; black offspring impossible.
3. a *Bb*
   b The brown mother and brown offspring were homozygous (*bb*). The black father must therefore have been heterozygous.
   c *B* and *b*
   d

		Sperm	
		B	b
Eggs	*b*	*Bb*	*bb*

   e Predict: 1/2 **f** Actual: 3/8 **g** It is a matter of chance which of the two kinds of sperm (*B* and *b*) fertilises each egg.
4. Testcross: to find whether an organism is homozygous or heterozygous.
5. Phenotype used: recessive

ISBN: 9780170372855

6 Each fertilisation is a random event, and is not predicted by whatever happens in other fertilisations. The predicted ratio of 25% brown mice may occur in a big sample, but not likely in a small sample of eight.

Activity B (page 117)

1 a I^OI^O
b I^O
c I^AI^B
d I^A or I^B
e

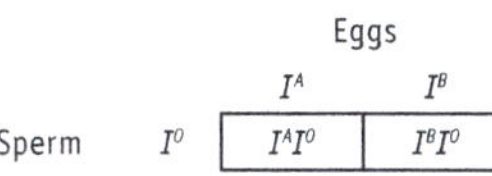

		Eggs I^A	Eggs I^B
Sperm	I^O	I^AI^O	I^BI^O

Since Joshua's gametes are all I^O, only two kinds of fertilisation event are possible; hence a two-cell Punnettt is the correct one to use.
f A: 50%. B: 50%. AB and O: impossible

2 a I^AI^A or I^AI^O
b I^AI^O
c Their baby (genotype I^OI^O) must have received an I^O allele from both parents, so both parents must have been heterozygous.
d

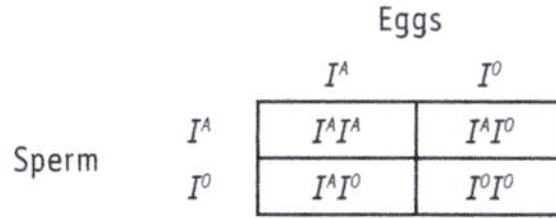

		Eggs I^A	Eggs I^O
Sperm	I^A	I^AI^A	I^AI^O
	I^O	I^AI^O	I^OI^O

e 75% chance, since three of the four different kinds of fertilisation event produce a group A child.
f Neither parent has the I^B allele.

3 Co-dominance: a situation where two alleles are expressed equally. Incomplete dominance: a situation where the heterozygote's phenotype is intermediate between the two homozygous phenotypes.

4 a The allele for yellow coats in mice, and the allele for taillessness in Manx cats.
b The ABO blood group alleles.
c The alleles for blood groups A and B.
d The alleles for colour in snapdragon flowers, and also in palomino horses.

5 Palomino horses are heterozygous *CW*, so when mated only half the offspring will be palomino.

		Eggs	
		C	*W*
Sperm	*C*	*CC* chestnut	*CW* palomino
	W	*CW* palomino	*WW* white

6

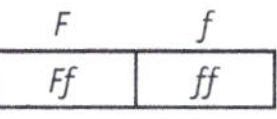

	F	*f*
F	*Ff*	*ff*

0% red: 50% white: 50% pink

7 a Both *Yy*
b 2 yellow : 1 grey
c At fertilisation this was probably a 1:2:1 ratio, but all *YY* embryos die in the uterus because the *Y* allele is lethal. Of the survivors, about 2/3 are yellow and 1/3 grey.

Activity C (page 121)

1 A trait is sex linked if it is controlled by a gene on the X chromosome. If it is recessive, it is more common in males, since they only have one X chromosome.

2 Red-green colour-blindness; also haemophilia.

3 a (1) The fact that it is only present in males is consistent with it being sex linked. (2) Both parents of #4 (and #10) have normal vision, showing that it is recessive. (3) #10 inherits it from his mother, whose X chromosome he has inherited.
b A mutation must have occurred in grandmother #1, since #4 must have inherited his X chromosome from her.
c #1 and #5; both have colour-blind sons, but both are normal.
d #7
e

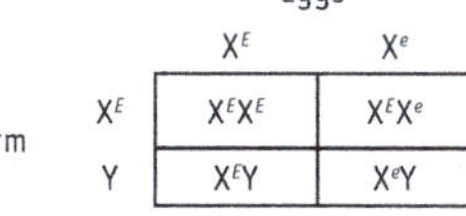

		Eggs X^E	Eggs X^e
Sperm	X^E	X^EX^E	X^EX^e
	Y	X^EY	X^eY

f Predicted: 1 0 1 1 1
g Actual: 1 affected boy, 2 normal boys, 1 girl carrier or normal
h It is a sheer chance as to which gametes meet

4 Colour-blind father, and a carrier (or colour-blind) mother.

5 The gene is on X chromosome. To be colour blind a boy need only inherit one colour-blind allele; a girl would need to inherit two, which is less likely.

6 a The trait skips a generation. Also, 6, 16 and 18 differ from their parents.
b It can't be sex linked as 18's father does not show the trait – so it must be autosomal.

Unit 4

Activity A (page 127)

1 Parent B genotype *rryy*

2 Parent B gametes *ry*

3 F_1 phenotypes: all round yellow

4 F_1 gametes *RY, Ry, ry, rY* (or in any other order; Punnettt below will still work)

5

	RY	*Ry*	*ry*	*rY*
RY	*RRYY*	*RRYy*	*RrYy*	*RrYY*
Ry	*RRYy*	*RRyy*	*Rryy*	*RrYy*
ry	*RrYy*	*Rryy*	*rryy*	*rrYy*
rY	*RrYY*	*RrYy*	*rrYy*	*rrYY*

6 F_2 expected ratios 9:3:3:1, as fractions of 16

7 Only *RRyy*

Activity B (page 128)

1 Black body and white face are dominant.

2 red body = *b*, black body = *B*, white face = *W*, black face = *w*

3 *BbWw*

4 *BW, Bw, bW, bw*

5 a *bbWw*
b *bW, bw*

6

	BW	*Bw*	*bW*	*bw*
bW	*BbWW*	*BbWw*	*bbWW*	*bbWw*
bw	*BbWw*	*Bbww*	*bbWw*	*bbww*

7 black body, white face = 3/8; black body, black face = 1/8; red body, white face = 3/8; red body, black face = 1/8

Activity C (page 130)

1 a tall red 25%, short red 25%, tall yellow 25%, short yellow 25%
b tall red 50%, short red 0%, tall yellow 0%, short yellow 50%
c The genes for stem height and flower colour are linked, i.e. the two loci must be on the same chromosome pair. The plants with new combinations of characteristics (tall, yellow-flowered, and short, red-flowered) result from crossing over. Since these recombinants are only 6% (3% + 3%), crossing over must be infrequent because the gene loci are close together.

2 a The locus of a gene is its position on the chromosome, relative to other genes.
b The two genes occupy different loci on the same chromosome 9.
c Blood group O, normal nails and patella.
d #6 is group A. Her mother (#4) was heterozygous (I^AI^O) for blood groups, but since #6 had inherited the NPS allele, she must also have inherited I^A allele since the two alleles are linked.
e If it were sex linked, #2 would have to pass the allele to all his daughters, but #3 is normal.
f Perhaps lethal in homozygotes.

3 *B–E–* = black, *bbee* = yellow, *B–ee* = red, *bbE–* = brown
a *bbee*
b black
c *BbEe* and *bbEe*
d *BBee*, red coat
e Parents must be: Male: *bbEe*, producing sperm *bE* and *be*
Female: *Bbee*, producing eggs *Be* and *be*.
Offspring possibilities:

		eggs	
		Be	*be*
sperm	*bE*	*BbEe* black	*bbEe* brown
	be	*Bbee* red	*bbee* yellow

Unit 5

Activity A (page 134)

1 number all-blue = 45 = 90% of population
number white-patch = 5 = 10% of population

2 24 *BB* genotype; so 48 *B* alleles
21 *Bb* genotype; so 21 *B* alleles, 21 *b* alleles
5 *bb* genotype; so 10 *b* alleles

3 Total *B* alleles = 48 + 21 = 69 out of 100, i.e. 69%; *f* = 0.69
Total *b* alleles = 21 + 10 = 31 out of 100, i.e. 31%; *f* = 0.31

4 Many of the all-blue birds are heterozygous, carrying the *b* allele for white wings, but this is not visible in the phenotype.

Activity B (page 138)

1

a	population
b	gene pool
c	frequency
d	founder effect
e	genetic bottleneck
f	cheetah
g	tieke
h	grey
i	disruptive
j	selection

2 a increase b decrease
c decrease d decrease
e increase f decrease
g increase h decrease
i decrease j decrease

ISBN: 9780170372855

3 Genetic bottlenecks and the founder effect both involve populations increasing from very low levels. NZ examples of bottlenecks: lesser spotted kiwi, saddleback (tieke). NZ examples of founder effect: pukeko, hares and many other introduced species.
4 Genetic bottlenecks involve an existing population crashing to near-extinction then recovering; founder effects involve a new population developing from a few individuals.
5 Selection in this case will result in the average length of flamingo legs becoming less. Graph will look like the one provided for albatross wings becoming shorter.
6 Black robins are not identical to that original 'bottleneck' female because (a) several males contributed to the bottleneck gene pool; (b) sexual reproduction with meiosis and fertilisation means that offspring are seldom identical anyway; (c) perhaps new mutations in the past 40 years.

Activity C (page 139)
1–4: results will vary.
5 Not very realistic because (a) the 'prey' were passive, not trying to escape or having defences other than colour; (b) prey did not improve with experience; (c) hunters did not improve or change behaviour with experience; (d) real hunters would probably persist for longer; (e) a wider range of colour backgrounds is likely; (f) unlikely there would be only three very different prey colours.

Revision 4

Genetic variation and change crossword (page 143)
Across: 1. locus; 6 heterozygous; 8 recessive; 9 trisomy; 12 homozygous; 14 fertilisation; 19 zygote; 20 discontinuous; 21 phenotype; 22 allele; 23 meiosis; 24 bottleneck.
Down: 2 chiasma; 3 stabilising; 4 mutation; 5 genetic; 7 crossing over; 10 autosome; 11 gene pool; 12 homologous; 13 genotype; 15 testcross; 16 directional; 17 polyploid; 18 disruptive; 20 Down.

2.6 | Ecology

Unit 1

Activity A (page 155)
Answer will vary.

Activity B (page 156)

1

a	habitat
b	community
c	ecosystem
d	food web
e	organism
f	abiotic
g	zonation
h	soil moisture
i	stratification
j	tapeworm
k	competitors
l	succession

2 a abiotic **b** biotic **c** biotic **d** biotic **e** abiotic **f** biotic **g** biotic **h** abiotic **i** abiotic

Activity C (page 157)

Community	Advantages	Disadvantages
1 Natural forest	Very diverse, may show stratification	Harder to get to?
2 Rocky shore intertidal zone	Very diverse, may show zonation	Harder to get to? Low tide needed
3 Undisturbed weedy area	Easy access, manageable size	Not so much species diversity
4 Lawn	Easy access, manageable size	Little diversity
5 Pond	Easy access (?)	Species diversity varies
6 Another area	?	?

Unit 2

Activity A (page 159)
1 a grid
b transect
c any (but return to the same areas each month)
d transect
e grid

2 a 120 cockles per square metre
b 1.2 million per hectare
c Both calculations assume (mistakenly?) that cockles are more or less evenly spread across the habitat, and that five samples each of 1/25 m^2 was sufficient.

Activity B (page 161)
1 Method A. Reason: the aim is to investigate plant damage, not count the possums themselves.
2 Method B at night (also trained sniffer dogs by day). Reason: kiwi are strictly nocturnal.
3 Method H. Reason: the area is too big to cover quickly in any other way. Also, elephants move across wide areas, so ground searches could result in double counting.
4 Method I. Reason: the area is manageably small; barnacles don't move.
5 Methods A, B, C, D.
6 Methods A and possibly B by day, D at night.
7 Method F.
8 Method G.

Activity C (page 162)
1 (Results will vary.)
2 (Answers will vary.)
3 Red trees are mostly randomly distributed, but some are associated with black trees.
Blue trees seem to be randomly distributed and isolated from each other.
Green trees tend to cluster in the northwest corner of the forest and are also spread randomly elsewhere.
Black trees occur mainly in two groups, and perhaps partly associated with red trees.
4 A regular grid system would be the best to detect the patterns described above.

2.7 | Gene expression

Unit 1

Activity A (page 166)
1 Keratin, function: prey capture, digging, etc., location: in fingernails, etc.
Collagen, function: attaching bones, location: in ligaments.
Insulin, function: controlling blood sugar, location: from pancreas.
Haemoglobin, function: carrying oxygen, location: in red blood cells.
Myosin, function: contracting, movement, location: in muscle.

2

a	polymer
b	polypeptide
c	monomer
d	peptide bond
e	tripeptide
f	keratin
g	condensation
h	haemoglobin
i	hydrolysis
j	myosin

3 20 different kinds of amino acids in proteins; one example is alanine.
4 Five: 20^5 = 3.2 million. Seven: 20^7 = 1.28 billion.

Activity B (page 169)
1 Primary structure of a protein: the sequence of amino acids along a polypeptide chain.
2 Secondary structure of a protein: the geometrically regular folding of the polypeptide chain into a helix or zig-zag.
3 The three-dimensional arrangement of R groups, which in turn is determined by the amino acid sequence.
4 Fibrous: keratin and collagen. Globular: haemoglobin
5 A protein is denatured when its tertiary structure is permanently destroyed, usually by heat.
6 Most of the forces holding a globular protein in its tertiary structure are weak hydrogen bonds, which are easily broken by heat.
7 The active site of an enzyme is the part of its surface that combines with the substrate and catalyses the reaction. It consists of amino acid R groups arranged in a specific three-dimensional pattern, held in place by the weak forces stabilising the tertiary structure of the enzyme.

Unit 2

Activity A (page 170)
1 chromosomes
2 nucleus
3 helix
4 nucleotides
5 deoxyribose sugar
6 hydrogen
7 adenine
8 and **9** cytosine and guanine
10 mitotic (cell)
11 copies/duplicates
12 complementary
13 23
14 G C A T C G A C C T A G C A T
C G T A G C T G G A T C G T A

ISBN: 9780170372855

Activity B (page 173)

1 As regards DNA length, a gene is a section responsible for synthesis of a particular polypeptide.
2 Transcription is a process in which a section of DNA is copied in mRNA language.
3 Translation is a process in which the information in mRNA governs the making of a polypeptide.
4 The role of RNA polymerase is to link together the nucleotides that make up RNA.
5 A codon consists of three successive mRNA nucleotides, associated with one particular amino acid.
6 The base sequences of the three 'stop' codons are UAA, UAG, UGA.
7

Feature	DNA	mRNA
Strand?	double	single
Sugar component?	deoxyribose	ribose
Comparative length?	long	shorter
Nucleotide bases?	ATCG	AUCG
Function? (< 15 words)	Retain coded information in the chromosomes, and pass it on to the next generation.	Uses copied information to govern the production of specific proteins.

Activity C (page 177)

1	DNA:	GGT	GTA	TTC	ACA	CAC		
	mRNA:	CCA	CAU	AAG	UGU	GUG		
	tRNA:	GGU	GTA	UUC	ACA	CAC		
	amino acids:	pro	his	lys	cys	val		
2	DNA:	TAC	ACC	CCA	CCG	CCT	AGC	ACT
	mRNA:	AUG	UGG	GGU	GGC	GGA	UCG	UGA
	amino acids:	start	trp	gly	gly	gly	ser	stop

3 transcription in nucleus; translation in cytoplasm/ribosomes
4

a	ribosomes
b	anticodon
c	codon
d	triplet
e	tRNA
f	mRNA
g	polymerase
h	peptide
i	transcription
j	translation

Unit 3

Activity A (page 178)

1	mRNA:	CAG	UAU	CCC	AAG
	amino acids:	gln	tyr	pro	lys
2	mutated mRNA:	CAU	AUC	CCA	AG
	amino acids:	his	ile	pro	–
3	mutated mRNA:	CAG	UGA	UUC	CAA
	amino acids:	gln	stop	–	–
4	mutated mRNA:	CAA	UAU	CCC	AAG
	amino acids:	gln	tyr	pro	lys

5 Substitution caused the least change because CAA and CAG both code for the same amino acid, and the other three codons are unaffected.
6 Insertion caused the greatest change because the second codon now signals 'stop', terminating the polypeptide.

Activity B (page 181)

1 Base insertion, base deletion.
2 Nonsense mutation: involves an entire sequence of amino acids changes. Mis-sense mutation: the replacement of one amino acid by another.
3 Benzene and formaldehyde.
4 The entire amino acid sequence 'downstream' of the mutation is changed, so the protein will be functionless.
5 A change in the third base will in many cases result in a triplet coding for the same amino acid.
6 UAG is a stop codon, so a mutation causing the stop codon will result: the polypeptide being shortened.
7 *Overall*: transcription is the first stage in which a gene-length section of DNA is rewritten as mRNA. Translation is the second stage in which info contained in mRNA is used to guide the final assembly of a polypeptide sequence. *Place*: transcription in nucleus; translation outside the nucleus in ribosomes. *Raw materials*: nucleotides for transcription; amino acids for translation. *Similarities*: both stages involve a flow of information, both are geared towards protein production.

Unit 4

Activity A (page 184)

1

a	albino	f	expression
b	tyrosine	g	PKU
c	precursor	h	PAH
d	metabolism	i	catalyse
e	tyrosinase	j	melanin

2 A metabolic pathway is a sequence of reactions in which one substance is converted, one step at a time, into another. Each step is catalysed by a particular enzyme.
3 a *JJKK* health normal; skin normal
b *JJKk* health normal; skin normal
c *JJkk* health normal; skin albino
d *JjKK* health normal; skin normal
e *JjKk* health normal; skin normal
f *Jjkk* health normal; skin albino
g *jjKK* health PKU; skin pale
h *jjKk* health PKU; skin pale
i *jjkk* health PKU; skin albino
j George *jjKK*, Mary *JJKK*.
k George probably inherited this condition from a mutation affecting the gametes of one of his parents, that parent possibly being heterozygous *Jj* to begin with.
4 a Enzymes
b Three genes, because three enzymes involved.
c Enzyme 2 would not work, so substances W and X would accumulate and Y and Z would decrease.
d Direct cause: enzyme 3 does not function; indirect: a mutation in the gene coding for this enzyme.
e It could be either the product of another metabolic pathway or a raw material from the environment.
5 a purple
b purple
c white
d purple
e purple
f white
g white
h white
i white
j *PQ*, *Pq*, *pQ* and *pq*
6

	PQ	*Pq*	*pq*	*pQ*
PQ	*PPQQ*	*PPQq*	*PpQq*	*PpQQ*
Pq	*PPQq*	*PPqq*	*Ppqq*	*PpQq*
pq	*PpQq*	*Ppqq*	*ppqq*	*ppQq*
pQ	*PpQQ*	*PpQq*	*ppQq*	*ppQQ*

7 9/16 purple:7/16 white

Activity B (page 188)

1 The expression of a gene is the characteristic that results after the gene has been translated into a protein. The 'expression' of a gene can be described in terms of the protein produced, or the effect that protein has on the rest of the phenotype.
2 The proportion of males would increase.
3 a In humans, the effect of UV light on skin is to stimulate production of the pigment melanin; and the adaptive value of this is to protect the DNA in skin cells against the mutagenic effects of UV.
b In arctic foxes, the effect of decreasing day length is causing the production of white, winter coat; and the adaptive value of this is winter white camouflage may help it escape predators and increase hunting success.
4 a Both groups no change in colour, as it takes more than two weeks for new fur to grow.
b Group 1: brown. Their skin is warm so tyrosinase production and melanin production both work normally. Group 2: cream colour. Skin is cold, so little melanin produced.
5 a Flower colour must be genetically determined, since different soil environments result in the same blue flower colour.
b In this case, flower colour would be influenced by soil type.
6 In these cats, one of the enzymes (tyrosinase) needed for producing melanin is very heat sensitive, and is denatuned even at normal body temperature. Result: little melanin is produced in warmer parts of the body. More melanin is produced in cooler parts such as ears and paws and tail, so these parts are dark coloured.

Revision 6

Gene expression crossword (page 193)

Across: 6 polypeptide; 10 hydrogen; 11 amino acid; 12 triplet; 13 quaternary; 14 nonsense; 19 mis-sense; 20 fibrous; 21 globular; 22 water; 23 polymer.
Down: 1 transcription; 2 degenerate; 3 complementary; 4 tertiary; 5 ribosome; 7 frameshift; 8 primary; 9 hydrolysis; 12 transfer RNA; 15 somatic; 16 messenger RNA; 17 point; 18 carboxyl; 19 mutagen.

2.8 | Microscope skills

Unit 1

Activity A (page 205)

1 eyepiece lens
2 coarse focus control
3 fine focus control
4 stage
5 base
6 mirror
7 revolving turret
8 high-power objective lens
9 low-power objective lens
10 stage clips
11 condenser lens and diaphragm

ISBN: 9780170372855

Activity B (page 206)

1

	Part	Function or description
a	Only the fine focus control	is to be used with high-power lenses.
b	The turret is	the rotating part holding three or four lenses.
c	An eyepiece is	the lens closest to your eye.
d	An objective lens is	the one closest to the slide.
e	Stage clips are used for	holding the slide in place.
f	The diaphragm is used for	regulating the amount of light.
g	The stage is	the platform on which slides are placed.

2

	Rule	Reason
a	When not in use, always put a plastic cover over a microscope.	To protect from dust, which is very harmful to optical instruments.
b	If the microscope has a mirror, never use it to reflect sunlight onto the slide.	Sunlight reflected into your eye could blind you.
c	Always look from the side when moving the HP objective lens slowly into place.	If the HP lens touches the slide, it will become scratched and ruined.
d	Never let water or your fingers or any object touch a lens.	Any marks or scratches on a lens could ruin it. (Cleaning best done with special tissue.)
e	Never force a focus control.	Focus mechanisms are easily broken if forced.

Activity C (page 207)

5

Object	Size in mm	in µm
Diameter red blood cell	7/1000 mm	7
Diameter of HP field	0.4 mm	400
Leaf epidermis cell	0.1 mm	100
Chloroplast diameter	1/100 mm	10
Mitochondrion length	0.003 mm	3
Paramecium length	0.2 mm	200

6 a 112 mm, 0.56 mm, 560 µm
b 4 mm, 20 µm
c 9 mm, 45 µm
d Cells at Z have very thick walls.
e Possible function: to support the plant stem, or to protect inner cells.

Unit 2

Activity A (page 211)

1

Cell or tissue feature	Function
Many chloroplasts	Absorb sunlight
Epidermis transparent	Maximise photosynthesis
Stomata	CO_2 entry for photosynthesis
Epidermis cells thick-walled	Reduce water loss
Long thick-walled cells with openings at the ends	Carry water in xylem cells

2

a	iodine
b	cavity slide
c	coverslip
d	toluidine
e	chloroplasts
f	epidermis
g	methyl cellulose
h	micrometre

Activity B (page 213)

1 Errors and omissions:
- most cells drawn not touching their neighbours and with large air gaps inbetween cells: not realistic
- many cells drawn carelessly, with cell walls showing gaps and overlapping lines
- little differentiation of cell shape
- cell walls are shown all the same thickness — not realistic
- chloroplasts carelessly drawn in only two cells, and not labelled
- cuticle indicator lines shown crossing other lines
- no title, no scale, no magnification
- specialised features of cell structure not noted, and no reference to the link between form and function.

2 Errors and omissions (this drawing contains many good features):
- only a few specialised features of cell structure are noted, but no reference to the link between form and function
- no title, no scale, no magnification given
- 'palisade' incorrectly spelled
- cell walls are shown all the same thickness — not realistic
- all lines a bit fuzzy; sharper pencil needed.

ISBN: 9780170372855